D0936874

# THE EVOLUTION OF ELECTRIC BATTERIES IN RESPONSE TO INDUSTRIAL NEEDS

by

SAMUEL RUBEN

DORRANCE & COMPANY • *Philadelphia and Ardmore, Pa.*

ISBN 0-8059-2455-8
Printed in the United States of America

*Dedicated to the late Philip Rogers Mallory, whose encouragement in difficult times was a source of inspiration.*

# CONTENTS

# ILLUSTRATIONS

# TABLES AND OPERATING DATA

# FOREWORD

We sometimes overemphasize the worthy contributions of the individuals responsible for technological advances but fail to recognize the motivating influence of industry. In reviewing the evolution of electric battery technology, we see a striking example of the importance of new industries in the development of new concepts. Industrial needs have provided the inspiring atmosphere for contemplation of invention and a market by means of which conceptual work could be translated into a working device or system. Thus the development and technological evolution of the electric battery clearly illustrate the relation of industrial needs to the related end product of research and development.

The opinion is held by some that many inventions which are developed or acquired by large organizations and which could be beneficially utilized by the public are held back or buried for fear of displacing or affecting the sale of other products. In my more than fifty years of independent research and invention, I have come in contact with the top management of many of the important manufacturers in the electrochemical and electronic fields, and I have not had experiences to support that opinion. The industrialist Philip Rogers Mallory, founder of P.R. Mallory & Co. Inc., had occasion to say more than once, "It is bad business to try to block progress." The success of the Mallory Company in developing and commercializing inventions, whether originating within the company or without, proves his point.

It is quite true that the N.I.H. (Not Invented Here) factor present in some engineering divisions has at times held back or blocked the acceptance of new ideas or inventions from the outside, but the management of every company I have licensed has always been anxious to bring a new product into the field and has, when necessary, overruled the lower echelons.

I mention the Mallory Company because it was a pioneer in the introduction of sealed alkaline primary cells.

This essay will describe the emerging needs for the industrial electric battery and their relation to evolving battery technology in the following order:

a. Early history and conceptual thinking
b. Initiation of electrodeposition of metals as a viable process in the production of early batteries
c. Telegraphy battery needs
d. Railroad battery needs for switch and signal circuitry
e. Telephone battery needs
f. Flashlight battery needs
g. Radio and portable communication battery needs
h. Household communication battery needs
i. Industries requiring miniaturized batteries, such as hearing aid, implanted cardiac pacemaker, calculators, and electric watches
j. Aerospace research requiring high-energy-density battery systems

Appreciation is acknowledged for the encouragement and constructive suggestions given by Professor Henry Linford. I wish also to express my appreciation to Dr. Bern Dibner, founder of the Burndy Library, for the use of Volta's letter to Sir Joseph Banks, translated from the French; to Leon Robbin and William Sauerbrey for the reading of the manuscript and helpful suggestions; to the Electrochemical Society for their permission to use parts of my paper, "Alkaline Primary Cells" (*Journal of the Electrochemical Society* Vol. 122, No. 7, July, 1975) on sealed zinc mercuric oxide cells, and to Joseph L. Dalfonso, president of Mallory Battery Company, a division of P.R. Mallory & Co. Inc., for his encouragement and assistance in providing data on Mallory batteries.

Samuel Ruben

# CHAPTER I

*The early observation of electrophysiological effects obtained by contact of dissimilar metals and moist muscle tissue—Swammerdam's observation in 1678 of muscular movement produced by contact with metals on nerves—Johann Sulzer's observation in 1762 of physiological effect of bimetal contact on the tongue—Galvani's frog's-leg experiments and recognition of muscular response to nerve contact by dissimilar metals.*

The reported experimental work of curiosity-inspired men led the way for scientists in the nineteenth century to conceive the practical means for generating electricity by chemical action in an electrochemical cell.

The electric battery is one of the great developments in the history of science. Its influence on the many disciplines of science initiated the advances of the nineteenth century that propelled us into our electricity-oriented society.

While the electric battery, or primary cell, is often thought of in terms of large-scale applications, such as the common flashlight and other applications requiring a portable source of electric energy, its role in the discovery of means for generating extensive industrial electrical power should also be recognized.

Following the discovery by Oersted of a magnetic field

surrounding a wire through which an electric current from a primary cell flowed, Ampere went on to discharge current through a copper-wire helix surrounding an iron rod, thus producing an electromagnet. These discoveries established the foundations of electrodynamics.

Faraday first applied these discoveries in establishing the principle of the electric motor, observing that a wire through which a current was discharged would rotate in a magnetic field. This led to the development by later investigators of the practical electric motor, thus reducing the burden of mechanical work by man.

According to Faraday's concept, the converse of producing a magnetic field by the discharge of electric current through a wire should produce an electric current by induction in a magnetic field. This led to his recognition of the potential generated by magnetic induction when he interrupted an electric circuit containing an electromagnet energized by the potential from a voltaic cell.

His next, and most important, discovery occurred when he rotated a copper disc between the poles of a magnet and found that an electric potential was developed between the shaft and the outer edge of the disc by magnetic induction. It was the primary voltaic cell that provided Faraday with electrical currents which made possible the greatest discovery in the history of technology, the electric generator for conversion of mechanical energy to electrical energy.

Today we take for granted the important inventions that were made during the Industrial Revolution of the eighteenth and nineteenth centuries, and those pioneering accomplishments tend to be forgotten with the dimmed vision of time.

In the early part of the eighteenth century a number of experimentalists reported physiological reactions induced by contact of dissimilar metals on a moist body organ without understanding their galvanic character.

Jon Swammerdam, a Dutch naturalist, reported in *Biblia Naturae* (Vol. II, Leipzig, 1737) that if a muscle supported on a copper rod in a glass tube had a silver wire wound around an extending nerve, and if its free end contacted the copper, a muscular movement could be observed.

A Swiss chemist, Johann Sulzer, reported a metal-contact

physiological sensation in his paper, "Theory of Agreeable and Disagreeable Sensations" (Berlin, 1762). He observed that there was no distinguishable difference in the taste of various individual metals, but if he put two dissimilar metals in his mouth and they contacted each other, he experienced an acrid taste as they touched his tongue. He described this taste as being like that of iron sulfate. Unless the metal strips that he used, such as silver and lead wire, were in physical contact, there was no distinguishable taste. He explained the taste induced by bimetal contact as being caused by the difference in particle vibration of the metals on the nerves of the tongue. The electrochemical, or galvanic, effect was not understood at that time.

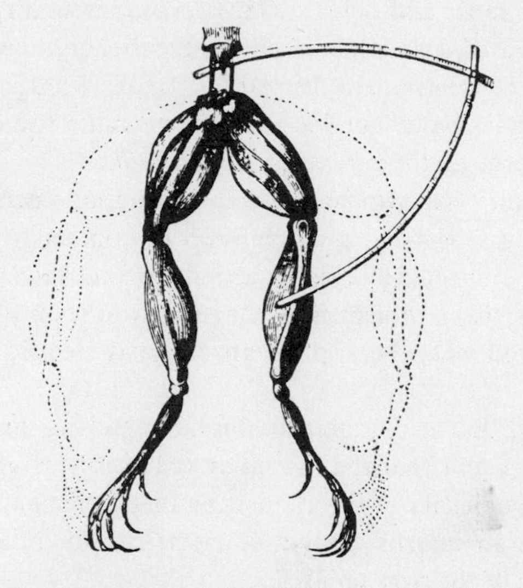

*Fig. 1. Frog's Legs (Galvani).* Galvani's frog's legs kicked when a copper rod touched the main nerve and an iron wire touched the rod and a leg nerve at the same time. Volta correctly found the reason for this startling action to be in a new form of electricity.

Luigi Galvani, a professor of anatomy at the University of Bologna, had been occupied in the study of the influence of electricity on the nervous system of animals, particularly frogs. Before 1800, the only means of obtaining electrical discharge was from electrostatic generators, which depend upon friction effects on the surface of insulating bodies. The most common type of electric machine utilized, as its potential generating agent, rotating glass discs in contact with friction pads connected to stationary terminals. This device would develop intermittent discharges of high potential when the discs were rotated. In order to provide limited storage for the electric force generated, the terminals were connected to a condenser, or capacitor, which stored the high potential in the form of electrical stress imposed upon glass plates or jars having metal foil on opposite sides; the latter served as charging surfaces when connected to a source of potential. The condensers used at the time were of the Leyden jar type, constructed of a glass jar with the inner and outer surfaces coated with a thin metal foil. While such a capacitor increased the total energy per discharge, it was limited to intermittent discharges. In his investigation, Galvani used charged Leyden condensers to determine the effect of electrical discharge on the nervous system of animals.

In an historic experiment Galvani found that when he suspended the legs of a frog from a copper rod in contact with the lumbar nerves and connected a bridging arc of iron between the copper rod and a leg nerve, a violent muscular reaction took place similar to that observed when he applied an external electric charge to the same area.

He found that it required dissimilar metals to produce the observed effect and that the muscular reaction was greater with increased dissimilarity of the contacting metal elements. Thus he did not require an external source of electrical potential to cause the contraction of the frog's muscle.

Over the years Galvani steadfastly held to his explanation that it was a flow of animal electricity to the contacting rod and wire that caused the motor response of the muscles. Galvani had begun his studies on electric discharge effects on the nervous system of animals a decade earlier, by noting the violent muscular reaction of fish that came in contact with the marine torpedo, a species of electric eel capable of emitting strong electrical discharges. These observations undoubtedly influenced his consistently held opinion that animal electricity provided the energy for the muscular contraction of the frog's legs.

Galvani published the results of his discoveries in the Proceedings of the Bologna Academy of Science in 1791 in a paper entitled "De Viribus Electricitatis di Motu Musculari." He had a dozen copies made and sent them to his colleagues, and many wrote of their interpretations of Galvani's discovery; among them Fourbes published a paper entitled "Experiments and Observations Relative to the Influence Lately Discovered by M. Galvani" (Edinburgh, 1793). A number of prominent physiologists agreed with Galvani's theory.

ALOYSII GALVANI

DE

VIRIBUS ELECTRICITATIS

IN

MOTU MUSCULARI.

COMMENTARIUS.

BONONIÆ

Ex Typographia Instituti Scientiarum. 1791.

*CUM APPROBATIONE.*

Alessandro Volta, Professor of Physics at the University of Pavia, had received the paper with great interest. Being a physicist, he looked for an explanation based on the physical factors of generation of an electromotive force between dissimilar metals in contact with the moist tissue of the frog's leg. On May 5, 1792, Volta delivered a paper in the Hall of the University of Pavia in which he discussed Galvani's theory of muscular response to the flow of animal electricity and presented his own findings and theories, which were in conflict with those of Galvani.

The terms "galvanic" and "galvanism," used to describe the production of electricity by chemical action, are taken from Galvani's name, as is "galvanometer," the name given to an instrument for detecting the existence and determining the strength of small electric currents.

# CHAPTER II

*The Volta cell. Volta's letter to Sir Joseph Banks—Basic limitation of cathodic polarization with continuous electric current flow—Reduction of polarization by use of large-area electrodes, initiated by Cruikshank; structural changes by Pepys and Hare for supporting very large-area electrodes—Recognition by Davy of the chemistry of the Volta cell—Application of oxidizing electrolytes in contact with cathode to reduce hydrogen polarization by dual electrolytes, achieved by Davy, Grove, Bunsen, Poggendorff. Application of porous cup electrolyte separators—Mechanical means developed by Wollaston for removing electrodes from cell electrolyte when cell is not in use. Plunge-type cell of Pepys, Hare, and Grenet for removing electrode from contact with electrolyte when not in use—Reduction of local zinc corrosion by mercury amalgamation of anode, as reported by Kemp and others.*

Volta had observed the reactions reported by Galvani of more energetic contractions of a frog's leg when the nerve-contacting metal arc was composed of metal dissimilar to that of a supporting rod. Rather than believing that the response of the frog's muscle was caused by an electric flow of current generated by animal electricity, he attributed the reaction to energy supplied to the responsive nerves by a potential generated by contact of dissimilar metals in a closed circuit.

He stated that the disengagement of electricity from first-class metallic conductors to second-class nonmetallic conductors, such as moist tissues, was the source of the activating potential. The bimetal arc in a closed circuit with the moist frog's leg separating the metals causes a muscular response to the extent of the galvanic

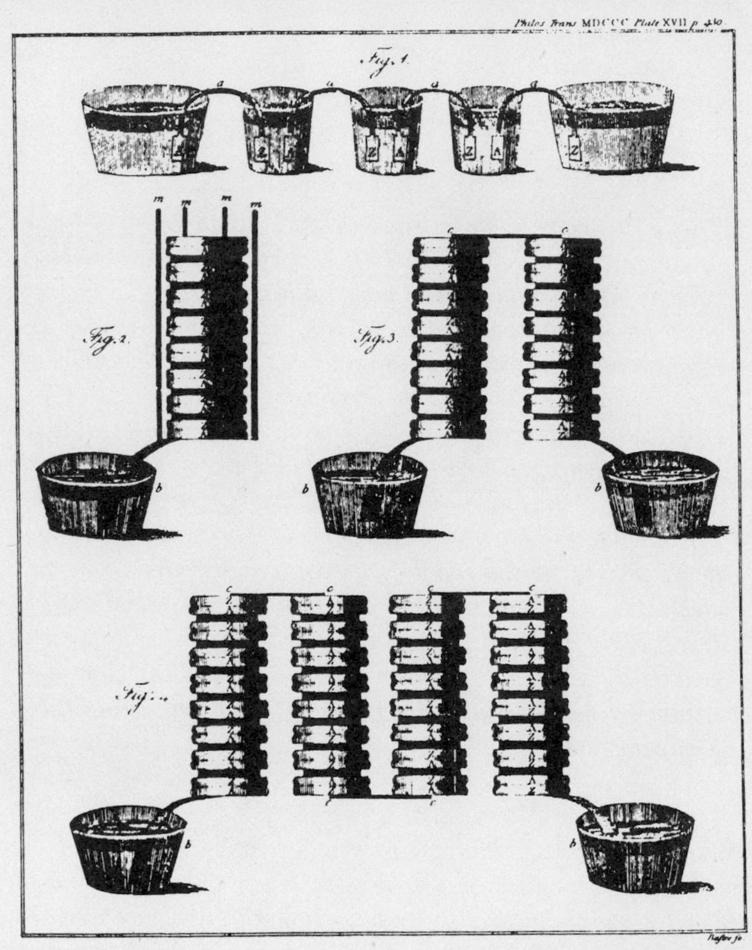


*Fig. 2. Volta Pile and Cells.* The illustration in Volta's letter to Banks shows four variations of the new electric battery. At the top is shown the "crown of cups" and below it variant arrangements of A (silver, for argentum) and Z, zinc discs with moist paper separators.

potential generated. This muscular reaction of the electrically charged nervous system serves to provide a sensitive electroscope that can indicate, by comparative movement, the difference in potential generated by various bimetal couples.

Volta propounded his contact theory with the principle that when two heterogeneous substances are in contact, one of them assumes a positive, and the other a negative, charge. His experimental work proved to be in the correct direction and led to his discovery of the means for generating a continuous flow of electricity by chemical action from dissimilar metals separated by an electrolyte.

Volta had been engaged in the study of his electric pile a number of years before he presented his discovery to the Royal Society. In announcing his epochal discovery of the generation of a continuous flow of electricity, he presented his theories and the structures that reflected the state of galvanic technology of that period.

Rather than give a hindsight description of Volta's cell or pile, I reproduce Volta's informative report to Sir Joseph Banks, president of the Royal Society in London, March 20, 1800.

> "On the Electricity excited by the mere Contact of conducting Substances of different Kinds." In a Letter from Mr. Alexander Volta, F.R.S. Professor of Natural Philosophy in the University of Pavia, to the Right Hon. Sir Joseph Banks, Bart. K.B.P.R.S.*
>
> Como in the Milanese, March 20, 1800.

> After a long silence, for which I shall offer no apology, I have the pleasure of communicating to you, and through you to the Royal Society, some striking results I have obtained in pursuing my experiments on electricity excited by the mere mutual contact of different kinds of metal, and even by that of other conductors, also different from each other, either liquid or containing some liquid, to which they are properly indebted for their conducting power. The principal of these results, which comprehends nearly all the rest, is

*Translated from Volta's paper published in French in the *Philosophical Transactions* of the Royal Society for 1800, pt. 2. Courtesy of Burndy Library, Norwalk, Conn.

the construction of an apparatus having a resemblance in its effects (that is to say, in the shock it is capable of making the arms, &c. experience) to the Leyden flask, or, rather, to an electric battery weakly charged acting incessantly, which should charge itself after each explosion; and, in a word, which should have an inexhaustible charge, a perpetual action or impulse on the electric fluid; but which differs from it essentially both by this continual action, which is peculiar to it; and because, instead of consisting, like the common electric jars and batteries, of one or more insulating plates or thin strata of those bodies which are alone thought to be *electric,* armed with conductors, or bodies called *non-electric,* this new apparatus is formed merely of several of the latter bodies, chosen from among those which are the best conductors, and therefore the most remote, as has hitherto been believed, from the electric nature. The apparatus to which I allude, and which will, no doubt, astonish you, is only the assemblage of a number of good conductors of different kinds arranged in a certain manner. Thirty, forty, sixty, or more pieces of copper, or rather silver, applied each to a piece of tin, or zinc, which is much better, and as many strata of water, or any other liquid which may be a better conductor, such as salt water, ley, &c. or pieces of pasteboard, skin, &c. well soaked in these liquids; such strata interposed between every pair or combination of two different metals in an alternate series, and always in the same order of these three kinds of conductors, are all that is necessary for constituting my new instrument, which, as I have said, imitates the effects of the Leyden flask, or of electric batteries, by communicating the same shock as these do; but which, indeed, is far inferior to the activity of these batteries when highly charged, either in regard to the force and noise of the explosions, the spark, the distance at which the discharge may be effected, &c. as it equals only the effects of a battery very weakly charged, though of immense capacity; in other respects, however, it far surpasses the virtue and power of these batteries, as it has no need, like these, of being previously charged by means of foreign electricity, and as it is capable of giving a shock every time it is properly touched, however often it may be.

To this apparatus, much more similar at bottom, as I shall show, and even such as I have constructed it, in its form to the *natural electric organ* of the torpedo or electric eel, &c. than to the Leyden flask and electric batteries, I would wish to give the name of the *artificial electric organ:* and, indeed, is it not, like it, composed entirely of conducting bodies? Is it not also active of itself without any previous charge, without the aid of any electricity excited by any of the means hitherto known? Does it not act incessantly, and without intermission? And, in the last place, is it not capable of giving every moment shocks of greater or less strength, according to circumstances—shocks which are renewed by each new touch, and which, when thus repeated or continued for a certain time, produce the same torpor in the limbs as is occasioned by the torpedo, &c.?

I shall now give a more particular description of this apparatus and of others analogous to it, as well as of the most remarkable experiments made with them.

I provide a few dozens of small round plates or disks of copper, brass or rather silver, an inch in diameter more or less (pieces of coin for example), and an equal number of plates of tin, or, what is better, of zinc, nearly of the same size and figure. I make use of the term *nearly*, because great precision is not necessary, and the size in general, as well as the figure of the metallic pieces, is merely arbitrary: care only must be taken that they may be capable of being conveniently arranged one above the other, in the form of a column. I prepare also a pretty large number of circular pieces of pasteboard, or any other spongy matter capable of imbibing and retaining a great deal of water or moisture, with which they must be well impregnated in order to insure success to the experiments. These circular pieces of pasteboard, which I shall call moistened disks, I make a little smaller than the plates of metal, in order that, when interposed between them, as I shall hereafter describe, they may not project beyond them.

Having all these pieces ready in a good state, that is to say, the metallic disks very clean and dry, and the non-metallic ones well moistened with common water, or, what is much better, salt water, and slightly wiped that the moisture may not drop off, I have nothing to do but to arrange them, a matter exceedingly simple and easy.

I place then horizontally, on a table or any other stand, one of the metallic pieces, for example one of silver, and over the first I adapt one of zinc; on the second I place one of the moistened disks, then another plate of silver followed immediately by another of zinc, over which I place another of the moistened disks. In this manner I continue coupling a plate of silver with one of zinc, and always in the same order, that is to say, the silver below and the zinc above it, or vice versa, according as I have begun, and interpose between each of these couples a moistened disk. I continue to form, of several of these stories, a column as high as possible without any danger of its falling.

But, if it contain about twenty of these stories or couples of metal, it will be capable not only of emitting signs of electricity by

Cavallo's electrometer, assisted by a condenser, beyond ten or fifteen degrees, and of charging his condenser by mere contact so as to make it emit a spark, &c. but of giving to the fingers with which its extremities (the bottom and top of the column) have been touched several small shocks, more or less frequent, according as the touching has been repeated. Each of these shocks has a perfect resemblance to that slight shock experienced from a Leyden flask weakly charged, or a battery still more weakly charged, or a torpedo in an exceedingly languishing state, which imitates still better the effects of my apparatus by the series of repeated shocks which it can continually communicate.

To obtain such slight shocks from this apparatus which I have described, and which is still too small for great effects, it is necessary that the fingers, with which the two extremities are to be touched at the same time, should be dipped in water, so that the skin, which otherwise is not a good conductor, may be well moistened. To succeed with more certainty, and receive stronger shocks, a communication must be made, by means of a metallic plate sufficiently large, or a large metallic wire, between the bottom of the column (that is to say, the lower piece of metal,) and water contained in a bason or large cup, in which one, two, or three fingers or the whole hand is to be immersed, while you touch the top or upper extremity (the uppermost or one of the uppermost plates of the column) with the clean extremity of another metallic plate held in the other hand, which must be very moist, and embrace a large surface of the plate held very fast. By proceeding in this manner, I can obtain a small pricking or slight shock in one or two articulations of a finger immersed in the water of the bason, by touching, with the plate grasped in the other hand, the fourth or even third pair of metallic pieces. By touching then the fifth, the sixth, and the rest in succession till I come to the last, which forms the head of the column, it is curious to observe how the shocks gradually increase in force. But this force is such, that I receive from a column formed of twenty pairs of pieces (not more) shocks which affect the whole finger with considerable pain if it be immersed alone in the water of the bason; which extend (without pain) as far as the wrist, and even to the elbow, if the whole hand, or the greater part of it, be immersed; and are felt also in the wrist of the other hand.

I still suppose that all the necessary attention has been employed in the construction of the column, and that each pair or couple of metallic pieces, resulting from a plate of silver applied over one of zinc, is in communication with the following couple by a sufficient stratum of moisture, consisting of salt water rather than common water, or by a piece of pasteboard, skin, or any thing of the same kind well impregnated with this salt water. The disk must not be too small, and its surface must adhere closely to those of the metallic plates between which it is placed. This exact and extensive application of moistened disks is very important, whereas the metallic plates of each pair may only touch each other in a few points, provided that their contact is immediate.

All this shows that, if the contact of the metals with each other in some points only be sufficient (as they are excellent conductors) to give a free passage to a moderately strong current of electricity, the case is not the same with liquids, or bodies impregnated with moisture, which are conductors much less perfect; and which, consequently, have need of more ample contact with metallic conductors, and still more with each other, in order that the electric fluid may easily pass, and that it may not be too much retarded in its course, especially when it is moved with very little force, as in the present case.

In a word, the effects of my apparatus, that is to say, the shocks felt, are considerably more sensible in proportion as the temperature of the ambient air, or that of the water or moistened disks which enter into the composition of the column, and that of the water even in the bason, is warmer, as heat renders the water a better conductor. But almost all the salts, and particularly common salt, will render it a still better. This is one of the reasons, if not the only one, why it is so advantageous that the water of the bason, and, above all, that interposed between each pair of metallic plates, as well as the water with which the circular pieces of pasteboard are impregnated, &c. should be salt water, as already observed.

But all these means and all these attentions have only a limited advantage, and will never occasion your receiving very strong shocks as long as the apparatus consists but of one column formed only of twenty pairs of plates, even though they may consist of the two metals properest for these experiments, viz. silver and zinc; for if they were silver and lead, or tin, or copper and tin, the half of the effect would not be produced, unless the weaker effect of each pair were supplied by a much greater number. What really increases the electric power of this apparatus, and to such a degree as to make it equal or surpass that of the torpedo or electric eel, is the number of plates arranged in such a manner, and with the attention before mentioned. If to the twenty pairs above described twenty or thirty others be added disposed in the same order, the shocks which may be communicated by a column lengthened in this manner will be much stronger, and extend to both arms as far as the shoulder; and especially of that, the hand of which has been immersed in the water; this hand, with the whole arm, will remain more or less benumbed, if by frequently renewing the touches these shocks be made to succeed each other rapidly, and without intermission. This will be the case if the whole hand, or the greater part of it be immersed in the water of the bason; but if only one finger be immersed, either wholly or in part, the shocks being almost entirely concentrated in it alone, will become so much the more painful, and so acute as to be scarcely supportable.

It may readily be conceived that this column, formed of forty or fifty couples of metals, which gives shocks more than moderate to both the arms of one person, is capable of giving sensible shocks also to several persons, holding each other by the hands (sufficiently moist) so as to form an uninterrupted chain.

I shall now return to the mechanical construction of my apparatus, which is susceptible of several variations, and describe not all those which I have invented or made, either on a small or a large scale, but only a few, which are either curious or useful, which exhibit some real advantage, as being easier or sooner constructed, and which are certain in their effects, or can be longer preserved in good order.

I shall begin by one which, uniting nearly all these advantages, differs most in its figure from the columnar apparatus above described, but which is attended with the inconvenience of being much more voluminous. This new apparatus, which I shall call a *couronne de tasses* (a chain of cups), is represented Plate VIII.

I dispose, therefore, a row of several basons or cups of any matter whatever, except metal, such as wood, shell, earth, or rather glass (small tumblers or drinking glasses are the most convenient), half filled with pure water, or rather salt water or ley; they are made all to communicate by forming them into a sort of chain, by means of so many metallic arcs, one arm of which, Sa, or only the extremity S, immersed in one of the tumblers, is of copper or brass, or rather of copper plated silver; and the other, Za, immersed into the next tumbler, is of tin, or rather of zinc. I shall here observe, that ley and other alkaline liquors are preferable when one of the metals to be immersed is tin; salt water is preferable when it is zinc. The two metals of which each arc is composed, are soldered together in any part above that which is immersed in the liquor, and which must touch it with a surface sufficiently large; it is necessary therefore that this part should be a plate of an inch square, or very little less; the rest of the arc may be as much narrower as you choose, and even a simple metallic wire. It may also consist of a third metal different from the two immersed into the tumblers, since the action of the electric fluid which results from all the contacts of several metals that immediately succeed each other, or the force with which this fluid is at last impelled, is absolutely the same, or nearly so, as that which it would have received by immediate contact of the first metal with the last without any intermediate metals, as I have ascertained by direct experiments, of which I shall have occasion to speak hereafter.

A series of 30, 40 or 60 of these tumblers connected with each other in this manner, and ranged either in a straight or curved line, or bent in every manner possible, forms the whole of this new apparatus, which at bottom and in substance is the same as the other columnar one above described; as the essential part, which consists in the immediate communication of the different metals which form each couple, and the mediate communication of one couple with the other, viz. by the intervention of a humid conductor, exist in the one as well as the other.

In regard to the manner of trying these tumblers, and the different experiments for which they may be employed, there is no need of saying a great deal after the ample explanation I have already given respecting the columnar apparatus. It may be readily comprehended, that to obtain a shock it will be sufficient to immerse one hand into one of the tumblers, and a finger of the other hand into another of the tumblers at a considerable distance from the former; that this shock will be stronger the further these glasses are from each other; that is to say, in proportion to the number of the intermediate glasses, and consequently, that the strongest shock will be received when you touch the first and last end of the chain. It will be readily comprehended also, how and why the experiments will succeed much better by grasping and holding fast in one hand, well moistened, a pretty large plate of metal (in order that the communication may be more perfect, and formed in a great number of points), and touching with this plate the water in the tumbler, or rather the metallic arc, while the other is immersed in the other distant tumbler, or touches with a plate, grasped in the like manner, the arc of the latter. In a word, one may comprehend and even foresee the success of a great variety of experiments which may be made with this apparatus or chain of cups much more easily, and in a manner more evident, and which, if I may be allowed the expression, speak more to the eyes than those with the columnar apparatus. I shall therefore forbear from describing a great number of these experiments, which may be easily guessed, and shall relate only a few which are no less instructive than amusing.

Let three twenties of these tumblers be ranged, and connected with each other by metallic arcs, but in such manner, that, for the first twenty, these arcs shall be turned in the same direction; for example, the arm of silver turned to the left, and the arm of zinc to the right; and for the second twenty in a contrary direction, that is to say, the zinc to the left, and the silver to the right; in the last place, for the third twenty, the silver to the left, as is the case in regard the first. When every thing is thus arranged, immerse one finger in the water of the first tumbler, and, with the plate grasped in the other hand, as above directed, touch the first metallic arc (that which joins the first tumbler to the second), then the other arc which joins the second and third tumbler, and so on, in succession, till you have touched them all. If the water be very salt and luke-warm, and the skin of the hands well moistened and softened, you will already begin to feel a slight shock in the finger when you have touched the fourth or fifth arc (I have experienced it sometimes very distinctly by touching the third), and by successively proceeding to the sixth and the seventh, &c. the shocks will gradually increase in force to the twentieth arc, that is to say, to the last of those turned in the same direction; but by proceeding onwards to the 21st, 22d, 23d or 1st, 2d, 3d of the second twenty, in which they are all turned in a contrary direction, the shocks will each time become weaker, so that at the 36th or 37th, they will be imperceptible, and be entirely null at the 40th, beyond which (and beginning the third twenty, opposed to the second and analogous to the first) the shocks will be imperceptible to the 44th or 45th arc; but they will begin to become sensible, and to increase gradually, in proportion as you advance to the 60th, where they will have attained the same force as that of the 20th arc.

If the twenty arcs in the middle were all turned in the same direction as the preceding twenty and the following twenty, that is to say, if the whole 60 conspired to impel the electric fluid in the same direction, it may readily be comprehended how much greater the effect will be at the end, and how much stronger the shock; and it may be comprehended, in general, to what point it must be weakened in all cases where a greater or smaller number of these forces act contrary to each other by an inverted position of metals.

If the chain be in any part interrupted, either by one of the tumblers being empty of water, or one of the metallic arcs being removed or divided into two pieces, you will receive no shock when you immerse your finger into the water of the first and another into that of the last vessel; but you will have it strong or weak, according to circumstances (leaving these fingers immersed), at the moment when the interrupted communication is restored; at the moment when another person shall immerse into the two tumblers, where the arc is wanting, two of his fingers (which will also receive a slight shock), or rather, when he shall immerse the same arc which has been taken away, or any other; and in the case of the arc separated into two pieces, at the moment when these pieces are again brought into mutual contact (in which case the shock will be stronger than in any other); and, lastly, in the case of the empty tumbler, at the moment when water poured into it shall rise to the two metallic arms immersed in this cup which before were dry.

When the chain of cups is of sufficient length, and capable of giving a strong shock, you will experience one, though much weaker, even though you keep immersed two fingers or the two hands, in one bason of water of pretty large size, in which the first and last metallic arcs are made to terminate; provided that either of these hands thus immersed, or rather both of them, be kept respectively in contact, or nearly in contact, with these arcs, you will, I say, experience a shock at the moment when (the chain being interrupted in any part) the communication is restored, and the circle completed in any of the ways before mentioned. One might be surprised that in this circle the electric current having a free passage through an interrupted mass of water, that which fills the bason, should quit this good conductor to throw itself and pursue its course through the body of the person who holds his hands immersed in the same water, and thus to take a longer passage. But the surprise will cease if we reflect, that living and warm animal substances, and above all, their humours, are, in general, better conductors than water. As the body, then, of the person who immerses his hands in the water, affords an easier passage than this water does to the electric current, the latter must prefer it though a little longer. In a word, the electric fluid, when it must traverse imperfect conductors in a large quantity, and particularly moist conductors, has a propensity to extend itself in a larger stream, or to divide itself into several, and even to pursue a winding course, as it thereby finds less resistance than by following one single channel, though shorter; in the present case it is only a part of the electric current, which, leaving the water, pursues this new route through the body of the person, and traverses it from the one arm to the other; a greater or less part passes through the water in the vessel. This is the reason why the shock experienced is much weaker than when the electric current is not divided when the person alone forms the communication between one arc and another, &c.

From these experiments one might believe, that when the torpedo wishes to communicate a shock to the arms of a man or to animals which touch it, or which approach its body under the water (which shock is much weaker than what the fish can give out of the water), it has nothing to do but to bring together some of the parts of its electric organ in that place, where, by some interval, the communication is interrupted, to remove the interruptions from between the columns of which the said organ is formed, or from between its membranes in the form of thin disks, which lie one above the other from the bottom to the summit of each column; it has, I say, nothing to do but to remove these interruptions in one or more places, and to produce there the requisite contact, either by compressing these columns, or by making some moisture to flow in between the pellicles or diaphragms which have been separated, &c. This is what may be, and what I really conclude to be, the task of the torpedo when it gives a shock; for all the rest, the impulse and movement communicated to the electric fluid, is only a necessary effect of its singular organ, formed, as is seen, of a very numerous series of conductors, which I have every reason to believe sufficiently different from each other to be *exciters* of the electric fluid by their mutual contacts; and to suppose them ranged in a manner proper from impelling that fluid with a sufficient force from top to bottom, or from the bottom to the top, and for determining a current capable of producing the shock, &c. as soon and as often as all the necessary contacts and communications take place.

But let us now leave the torpedo, and its *natural electric organ,* and return to the *artificial electric organ* of my invention, and particularly to my first *columnar apparatus*, that which imitates the first even in its form (for that composed of tumblers is different in that respect). I might say something also in regard to the construction of the said apparatus with tumblers or a *chain of glasses;* for example, that the first and last tumbler should be of such a size that, when necessary, the whole hand might be immersed in it, &c.: but, to enter into all these details, would require too much time.

In regard to the columnar apparatus, I endeavoured to discover the means of lengthening it a great deal by multiplying the metallic plates in such a manner as not to tumble down; and I discovered, besides others, the following, which are represented in the annexed figures. (Plate VIII. fig. 2, 3, 4)

In Fig. 2, mmmm are rods, three, four, or more in number, which rise from the bottom of the column, and confine, as in a cage, the plates or disks, placed each above the other in such number and to such a height as you choose, and which thus prevent them from falling. The rods may be of glass, wood, or metal, only that, in the last case, you must prevent them from coming into contact with the plates; which may be done either by covering each of them with a glass tube, or interposing between them and the column a few stripes of wax cloth, oiled paper, or even plain paper, and, in a word, any other body that may either be a *cohibent* or a bad conductor; wood or paper will be sufficiently so for our purpose, provided only that they are not very damp or moist.

But the best expedient, when you wish to form an apparatus to consist of a great number of plates, above 60, 80, or 100 for example, is to divide the column into two or more, as seen Fig. 3 and 4, (Plate VIII), where the pieces all have their respective positions and communication as if there were only one column. Fig. 4, as well as Fig. 3 may indeed be considered as a bent column.

In all these figures the different metallic plates are denoted by the letters S and Z (which are the initials of silver and zinc); and the *moistened disks* (of pasteboard, skin, &c. interposed between each pair of metals), are represented by a black stratum. The plates of metal may either be laid simply upon each other and so brought into union in an indefinite number of points, or they may be soldered together. It is altogether indifferent which ever of these methods be followed. cc, cc, cc, are the metallic plates which form a communication between each column, or section of a column, and another; and bb, bb, bb, are the basons of water in communication with the lower part or extremities of these columns.

An apparatus thus prepared is exceedingly convenient without being bulky; and it might be rendered portable, with still more ease and safety, by means of circular cases or tubes, in which each column might be inclosed and preserved. It is only to be regretted that it does not long continue in a good state; the moistened disks become dry in one or two days to such a degree that they must be again moistened; which, however, may be done without taking to pieces the whole apparatus, by immersing the columns, completely formed, in water, and wiping them, when taken out some time after, with a cloth, or in any other manner.

The best method of making an instrument as durable as can be wished, would be, to inclose and confine the water interposed between each pair of metals, and to fix these metallic plates in their places by enveloping the whole column with wax or pitch; but this would be somewhat difficult in the execution, and would require a great deal of patience. I have, however, succeeded; and have formed in this manner two cylinders consisting of twenty pair of metals, which can still be employed though made several weeks, and which, I hope, will be serviceable for months.

These cylinders are attended with this advantage, that they may be employed for experiments either in an erect, inclined, or lying position, according as you choose, or even immersed in water, provided the top of it be above the surface of the fluid: they might also give a shock when entirely immersed if they contained a greater number of plates, or if several of these cylinders were joined together, and if there were any interruptions that could be removed at pleasure, &c. by which means these cylinders would have a pretty good resemblance to the electric eel; and, to have a better resemblance to it even externally, they might be joined together by pliable metallic wires or screw-springs, and then covered with a skin terminated by a head and tail properly formed, &c.

The effects sensible to our organs produced by an apparatus formed of 40 or 50 pairs of plates (and even by a smaller, if one of the metals be silver or copper and the other zinc) are reduced merely to shocks: the current of the electric fluid, impelled and excited by such a number and variety of different conductors, silver, zinc, and water, disposed alternately in the manner above described, excites not only contractions and spasms in the muscles, convulsions more or less violent in the limbs through which it passes in its course; but it irritates also the organs of taste, sight, hearing, and feeling, properly so called, and produces in them sensations peculiar to each.

And first, in regard to the sense of feeling: If, by means of an ample contact of the hand (well moistened) with a plate of metal, or rather, by immersing the hand to a considerable depth in the water of the bason, I establish on one side a good communication with one of the extremities of my *electro-motive apparatus,* (we must give new names to instruments that are new not only in their form, but in their effects or the principle on which they depend); and on the other I apply the forehead, eye-lid, tip of the nose, also well moistened, or any other part of the body where the skin is very delicate: if I apply, I say, with a little pressure, any one of these delicate parts, well moistened, to the point of a metallic wire, communicating properly with the other extremity of the said apparatus, I experience, at the moment that the conducting circle is completed, at the place of the skin touched, and a little beyond it, a blow and a prick, which suddenly passes, and is repeated as many times as the circle is interrupted and restored, so that, if these alternations be frequent, they occasion a very disagreeable quivering and prickering. But if all these communications continue without these alternations, without the least interruption of the circle, I feel nothing for some moments; afterwards, however, there begins at the part applied to the end of the wire, another sensation, which is a sharp pain (without shock), limited precisely by the points of contact, a quivering, not only continued, but which always goes on increasing to such a degree, that in a little time it becomes insupportable, and does not cease till the circle is interrupted.

What proof more evident of the continuation of the electric current as long as the communication of the conductors forming the circle is continued?—and that such a current is only suspended by interrupting that communication? This endless circulation of the electric fluid (this *perpetual motion*) may appear paradoxical and even inexplicable, but it is no less true and real, and you feel it, as I may say, with your hands. Another evident proof may be drawn from this circumstance, that in such experiments you often experience, at the moment when the circle is suddenly interrupted, a shock, a pricking, an agitation, according to circumstances, in the same manner as at the moment when it is completed; with this only difference, that these sensations, occasioned by a kind of reflux of the electric fluid, or by the shock which arises from the sudden suspension of its current, are of less strength. But I have no need, and this is not the place to bring forward proofs of such an endless circulation of the electric fluid in a circle of conductors, where there are some, which, by being a different kind, perform, by their mutual contact, the office of exciters or *movers:* this proposition which I advanced in my first researches and discoveries on the subject of galvanism, and always maintained by supporting them with new facts and experiments, will, I hope, meet with no opposers.

Recurring to the sensation of pain which is felt in the experiments above described, I must add, that if this pain be very strong and pricking in the parts covered by the skin, it is much more so in those where the skin has been taken off in recent wounds for example. If by chance there should be a small incision or bit of the skin rubbed off in the finger which I immerse in the water that communicates with one of the extremities of the *electro-motive* apparatus, I experience there a pain so acute, when, by establishing the proper communication with the other extremity, I complete the circle, that I must soon desist from the experiment; that is to say, must withdraw my finger, or interrupt the circle in some other manner. I will say more; that I cannot even endure it above a few seconds when the part of the apparatus which I put in play, or the whole apparatus, contains only twenty pair of plates, or about that number.

One thing, which I must still remark, is, that all these sensations of pricking and pain are stronger and sharper, every thing else being equal, when the part of the body which is to feel them is towards the negative electricity; that is to say, placed in such a manner in the conducting circle, that the electric fluid traversing that circle is not directed towards that sensible part, does not advance towards it, and enter from the outside inwards, but takes its direction from the inside outwards; in a word, that it issues from it: in regard to which it is necessary to know, of the two metals that enter by pairs into the construction of the machine, which is the one that gives off to the other. But I had already determined this respecting all the metals by other experiments, published a long time ago at the end of my first memoirs on galvanism. I shall therefore say nothing further here, than that the whole is completely confirmed by the experiments, equally and still more demonstrative and striking, with which I am at present employed.

In regard to the sense of taste, I had before discovered, and published in these first memoirs, where I found myself obliged to combat the pretended animal electricity of Galvani, and to declare it an external electricity moved by the mutual contact of metals of different kinds,—I had discovered, I say, in consequence of this power which I ascribed to metals, that two pieces of these different metals, and particularly one of silver and one of zinc, applied in a proper manner, excited at the tip of the tongue very sensible sensations of taste; that the taste was decidedly acid, if, the tip of the tongue being turned toward the zinc, the electric current proceeded against it, and entered it; and that another taste, less strong but more disagreeable, acrid, and inclining to alkaline, was felt, if (the position of the metals being reversed) the electric current issued from the tip of the tongue; that these sensations continued and received even an increase for several seconds, if the mutual contact of the two metals was maintained, and if the conducting circle was nowhere interrupted. But when I have said here, that exactly the same phenomena take place when you try, instead of one pair of these metallic pieces, an assemblage of several of them, ranged in the proper manner; and that the said sensations of taste, whether acid or alkaline, increase but a little with the number of these pairs, I have said the whole. It only remains for me to add that, if the apparatus put in play for these experiments on the tongue be formed of a sufficiently large number of metallic pairs of this kind, for example, if it contain 30, 40 or more, the tongue experiences not only the sensation of taste already mentioned, but, besides that, a blow which it receives at the moment when the circle is completed, and which occasions in it a pricking more or less painful, but fleeting, followed some moments after by a durable sensation of taste. This blow produces even a convulsion or agitation of a part of or the whole of the tongue, when the apparatus, formed of a still greater number of pairs of the said metals, is more active, and if, by means of good communicating conductors, the electric current which it excites be able to pass every where with perfect freedom.

I must often recur to, and insist on, this last condition, because it is essential in all experiments when you wish to obtain sensible effects on the body, or commotions in the limbs, or sensations in the organs of the senses. It is necessary, therefore, that the non-metallic conductors which enter into the circle should be as good conductors as possible, well moistened (if they are not themselves liquid) with water, or with any other liquid that may be a better conductor than pure water; and it is necessary, besides, that the well moistened surfaces by which they communicate with the metallic conductor, should be sufficiently large. The communication ought to be confined or reduced to a small number of points of contact only in that place where you wish to concentrate the electric action on one of the most sensible parts of the body, on any of the sensitive nerves, &c. as I have already remarked in speaking of the experiments on feeling, *viz.* those by which acute pains are excited in different parts. The best method which I have found for producing on the tongue all the sensations above described, is, to apply the tip of it to the pointed extremity (which, however, must not be too much so) of a metallic rod, which I made to communicate properly, as in the other experiments, with one of the extremities of my apparatus, and to establish a good communication between the hand, or, what is better, both the hands together, and the other extremity. This application of the tip of the tongue to the end of the metallic rod, may either exist already, when you are going to make the other communication to complete the circle (when you are going to immerse your hand into the water of the bason), or be made after the establishment of this communication, while the hand is immersed; and in the latter case I think I feel the pricking and shock in the tongue, a very short time before actual contact. Yes; it always appears to me, particularly if I advance the tip of my tongue gradually, that; when it has arrived within a very small distance of the metal, the electric fluid (I would almost say spark), overcoming this interval, darts forward to strike it.

In regard to the sense of sight, which I also found might be affected by the weak current of the electric fluid, arising from the mutual contact of two different metals in general, and in particular, of a piece of silver and one of zinc, it was natural to expect that the sensation of light, excited by my new apparatus, would be stronger in proportion as it contained a greater number of pieces of these metals; each pair of which, arranged in the proper manner, adds a degree of force to the said electric current, as all the other experiments show, and particularly those with the electrometer assisted by the condenser, which I have only mentioned and which I shall describe on another occasion. But I was surprised to find that with 10, 20, 30 pairs, and more, the flash produced neither appeared longer and more extended, nor much brighter than with one pair. It is true, however, that this sensation of weak and transient light, is excited by such an apparatus much easier and in different ways. To succeed, indeed, with one pair, the following are almost the only methods; *viz.* that one of the metallic pieces should be applied to the ball of the eye, or the eye-lid well moistened and that it should be made to touch the other metal applied to the other eye, or held in the mouth, which produces a flash much more beautiful; or, that this second metallic piece should be held in the moistened hand and then brought into contact with the former; or, in the last place that these two plates should be applied to certain parts of the inside of the mouth, making them communicate with each other. But with an apparatus of 20 or 30 pairs, &c. the same flash will be produced by applying the end of a metallic plate or rod placed in communication with one of the extremities of the apparatus, to the eye, while with one hand you form a proper communication with the other extremity; by bringing, I say, this plate into contact not only with the eye or any part of the mouth, but even the forehead, the nose, the cheeks, lips, chin, and even the throat; in a word, every part and point of the visage, which must only be well moistened before they are applied to the metallic plate. The form as well as the force of this transient light which is perceived varies a little, if the places of the face to which the action of the electric current is applied, be varied, if it be on the forehead, for example, this light is moderately bright, and appears like a luminous circle, under which figure it presents itself also in several other experiments.

But the most curious of all these experiments is, to hold the metallic plate between the lips, and in contact with the tip of the tongue, since when you afterwards complete the circle in the proper manner, you excite at once, if the apparatus be sufficiently large and in good order, and the electric current sufficiently strong and in good order, a sensation of light in the eyes, a convulsion in the lips, and even in the tongue, and a painful prick at the tip of it, followed by a sensation of taste.

I have now only to say a few words on hearing. This sense, which

I had in vain tried to excite with only two metallic plates, though the most active of all the exciters of electricity, *viz.* one of silver or gold, and the other of zinc I was at length able to affect it with my new apparatus, composed of 30 or 40 pairs of metals. I introduced, a considerable way into both ears, two probes or metallic rods with their ends rounded, and I made them to communicate immediately with both extremities of the apparatus. At the moment when the circle was thus completed I received a shock in the head, and some moments after (the communication continuing without any interruption) I began to hear a sound, or rather noise, in the ears, which I cannot well define; it was a kind of crackling with shocks, as if some paste or tenacious matter had been boiling. This noise continued incessantly, and without increasing, all the time that the circle was complete, &c. The disagreeable sensation, and which I apprehended might be dangerous, of the shock in the brain, prevented me from repeating this experiment.

There still remains the sense of smelling, which I have hitherto tried in vain with my apparatus. The electric fluid, which, when made to flow in a current in a complete circle of conductors, produces in the limbs and parts of the living body effects correspondent to their excitability, which stimulating in particular the organs or nerves of touch, taste, sight and hearing, excite in them some sensations peculiar to each of these senses, as I have found, produces in the interior of the nose only a pricking more or less painful, and commotions more or less extensive, according as the said current is weaker or stronger. And whence comes it, then, that it does not excite any sensation of smell, though, as appears, it stimulates the nerves of that sense? It cannot be said that the electric fluid of itself is not proper for producing odorous sensations, since, when it diffuses itself through the air in the form of aigrettes, &c. in the common experiments made with electric machines, it conveys to the nose a very sensible smell resembling that of phosphorus. Taking similitude into consideration, and reasoning from its analogy with other odoriferous matters, I will say, that it must completely diffuse itself throughout the air to excite smell; that it has need, like other effluvia, of the vehicle of the air to affect that sense in such a manner as to excite the sensations of smell. But in the experiments of which I speak, that is to say, of an electric current in a circle of conductors, all contiguous, and without the least interruption, this absolutely cannot take place.

All the facts which I have related in this long paper in regard to the action which the electric fluid excited, and when moved by my apparatus, exercises on the different parts of our body which the current attacks and passes through—an action which is not momentaneous, but which lasts, and is maintained during the whole time that this current can follow the chain not interrupted in its communications; in a word an action the effects of which vary according to the different degrees of excitability in the parts, as has been seen;—all these facts, sufficiently numerous, and others which may be still discovered by multiplying and varying the experiments of this kind, will open a very wide field for reflection, and of views, not only curious, but particularly interesting to medicine. There will be a great deal to occupy the anatomist, the physiologist, and the practitioner.

It is well known, by the anatomy which has been made of it, that the electric organ of the torpedo or electric eel, consists of several membranaceous columns, filled from one end to the other with a great number of plates or pellicles, in the form of very thin disks, placed one upon the other, or supported at very small distances by intervals, into which, as appears, some liquor flows. But we cannot suppose that any of these laminae are of an insulating nature, like glass, resin, silk, &c. and still less that they can either become electric by friction, or be disposed and charged in the same manner as the small Franklinian plates or small electrophores; nor even that they are sufficiently bad conductors to perform the office of a good and durable condenser, as Mr. Nicholson has supposed. The hypothesis of this learned and laborious philosopher, by which he makes of each pair of these pellicles, which he compares to leaves of talc, as many small *electrophores* or *condensers,* is indeed very ingenious, and is, perhaps, the best theory that has been devised to explain the phenomena of the torpedo, adhering to the hitherto known principles and laws of electricity. For, besides that the mechanism, by which, every time that the fish intended to give a shock, the respective separation of the plates on the whole or a great number of these electrophores or condensers ought to be effected all at once, and ought to establish on the one hand a communication between themselves of all the plates electrified *positively,* and on the other a communication between all those electrified *negatively,* as Mr. Nicholson supposes—besides, that this very complex mechanism appears too difficult, and a little agreeable to nature;—and besides, that the supposition of an electric charge originally impressed, and so durable in these pellicles performing the office of electrophores, is altogether gratuitous,—such a hypothesis falls entirely, since these pellicles of the organ of the torpedo are not, and cannot be, in any manner insulating or susceptible of a real electric charge, and much less capable of retaining it. Every animal substance, as long as it is fresh, surrounded with juices, and more or less succulent of itself, is a very good conductor. I say more, instead of being as cohibent as resins or talc, to leaves of which Mr. Nicholson has compared the pellicles in question, there is not, as I have assured myself, any living or fresh animal substance which is not a better *deferent* than water, except only grease and some oily humours. But neither these humours nor grease, especially semi-fluid or entirely fluid, as it is found in living animals, can receive an electric charge in the manner of insulating plates, and retain it; besides, we do not find that the pellicles and humours of the organ of the torpedo are greasy or oily. This organ therefore, composed entirely of conducting substances, cannot be compared either to the electrophore or condenser, or to the Leyden flask, or any machine excitable by friction or by any other means capable of electrifying insulating bodies, which before my discoveries were always believed to be the only ones originally electric.

To what electricity then, or to what instrument ought the organ of the torpedo or electric eel, &c. to be compared? To that which I have constructed according to the new principle of electricity, discovered by me some years ago, and which my successive experiments, particularly those with which I am at present engaged, have so well confirmed, *viz.* that conductors are also, in certain cases, exciters of electricity in the case of the mutual contact of those of different kinds, &c., in that apparatus which I have named the *artificial electric organ,* and which being at bottom the same as the natural organ of the torpedo, resembles it also in its form, as I have advanced.

Volta experimented with various metals to determine the maximum electromotive force developed between dissimilar metals. He found that the couple having the greatest potential difference was zinc and carbon, but it was not as practical to use in a pile as a couple of zinc and copper or silver. He later suggested an electromotive series of the elements, such as zinc, tin, lead, iron, copper, silver, gold, platinum and carbon, and noted that the further apart these elements were positioned in the series, the greater would be the electromotive force developed. Ritter independently devised a similar series at the same time. In later years this electrochemical series would be greatly expanded by the work of Faraday, Nernst, and others.

Volta considered his "pile" as a sort of self-charging condenser and similar in a way to the charging structure of the electric eel.

As described in his letter to Sir Joseph Banks, one means for indicating the electromotive force of his pile was by the shock he felt when he placed his hand in contact with the terminals of his battery. It is to be noted that in his drawing of his "couronne de Tasses" the two end cells have only one electrode, the electrolyte serving as terminal contacts. In 1782 he had invented the condensing electroscope which was an improvement over the ordinary gold-leaf instrument, since it could indicate lower potentials. It would serve him for electromotive force measurement of large stacks of cell structure.

The Volta Pile was not practical for obtaining a flow of electricity of a sufficient magnitude over a protracted time to fulfill the necessary requirements for commercial applications. One of the earliest applications of the Volta cell, or battery, was for the electroplating of metals from their solutions. However, commercialization of this application had to await the improvement of batteries by others and eventually the development of the electric generator.

The basic limitation in coulombic capacity of the Volta battery was related to the fact that it was in essence a half cell, since the generation of its electromotive force depended upon the oxidation of the zinc electrode ($Zn \rightarrow Zn^{++} + 2e^-$) serving as the anode; the cathode was composed of a copper or silver electrode that was essentially a collector for the hydrogen ions in the electrolyte, $2H^+ + 2e^- \rightarrow H_2$ being propelled to it by the potential it received from the load circuit connected to the zinc anode. The flow of current on closed circuit continued until the cathode became polarized.

The polarization of the Volta cell which limited the continuous flow of current was recognized by William Cruikshank, who, upon learning of the Volta battery from the report sent to Sir Joseph Banks, constructed batteries with large electrode areas that would proportionately extend the length of time during which the battery would provide a meaningful current flow before polarizing. In July 1800 (in the *Nicholson's Journal of Material, Philosophy, Chemistry, and the Arts* 4:187), he reported operation of a battery having sufficient current flow to deposit copper and silver from their chemical solutions.

Cruikshank's battery was composed of wide-area zinc and copper plate couples that were fitted in grooved resin-filled spacers in a wooden box. In this structure, each couple was sealed off and the space between the copper and zinc side of the separated couples was filled with electrolytes of either saline solutions or dilute acids. The term "cell" applied to batteries is said to have been derived from Cruikshank's cell-structure battery.

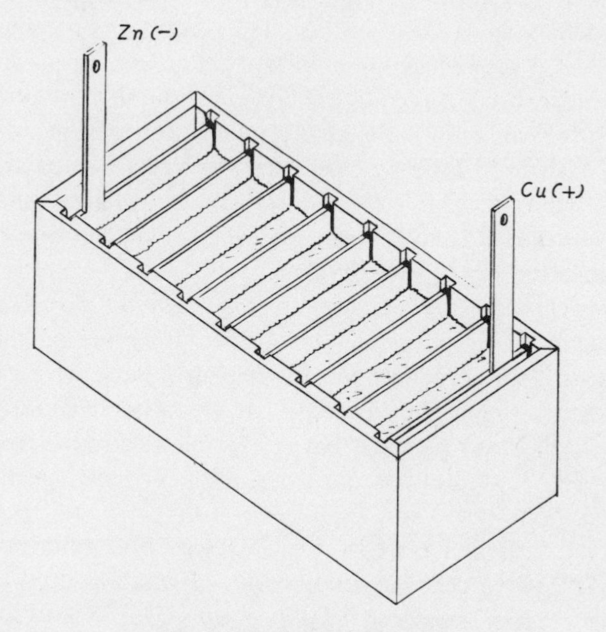


*Fig. 3. Cruikshank Battery*

In September 1800, Ritter published (in the *Annals of Physics* 6:468, Bavaria) a report of his work on the electroplating of metals with electric current derived from voltaic cells.

The early experiments of Cruikshank, Ritter, Brugnatelli, Berzelius, and others on electrodeposition initiated the development of an electroplating industry. However, a source of electric current was required that was not as limited in capacity by internal polarization, and the solution of this battery problem was initiated by chemists.

One of the earliest of these was Humphry Davy, who in September 1800 published a paper in the *Nicholson's Journal* entitled "On the Cause of Galvanic Phenomena" and followed this with a second paper in which he stated: "The oxidation of zinc in the pile and the chemical changes connected with it are somehow the cause of the electric effect it produces." He also found that the electromotive force and overall energy produced in a cell increased when he used an oxidizing acid, such as nitric, as the electrolyte.

Independently of Davy's work, J. Fabroni, a contemporary of Volta, had expressed his opinion that the oxidized appearance of the zinc discs used in a Volta Pile indicated the reaction on them to be the principal factor in the generation of electricity.

Van Marum in 1801 reported that the Volta Pile capacity was greatly increased if it were enclosed in an oxygen atmosphere. Pepys enclosed a Volta Pile in a nitrogen atmosphere and reported it had an effect opposite to oxygen.

However, others expanded on the work of Cruikshank in reducing polarization effects mechanically by having large-area electrodes. Pepys produced a large-plate-area battery for the Royal Institution with plates of 6 square feet area assembled in a battery of sixty cells. Davy used this battery for the electro-decomposition of halogen salt solutions and produced chlorine, bromine and fluorine.

In 1819 Professor Hare of the University of Pennsylvania produced large-area cells of spiral form by winding wide strips of zinc and copper separated by a fibrous spacer. These large-area electrodes in an acid electrolyte allowed high currents to be produced initially. Pepys improved Hare's spirally wound cell by converting it into a plunge-type structure in the manner of the Wollaston battery. This reduced electrode loss when the cell was not in use. The electrodes used by Pepys in his spirally wound, plunge-type Hare cell were strips of zinc and copper 50 feet long and 2 feet wide separated by horse-hair rope spacers. When activated by immersion into the acid, the cell proved capable of producing large currents that decreased with time on load to a low value determined by the extent of hydrogen accumulation on the cathode surface. Raising the electrodes out of the electrolyte would temporarily renew their capacity, since abrasion effects and contact with the oxygen in the atmosphere would remove the hydrogen film by diffusion and combination.

One of the outstanding discoveries that helped to make a more practical voltaic cell possible was the proposition by Kenneth T. Kemp *(Edinburgh New Philosophical Journal 6* [1828]: 70-77) that amalgamating zinc with mercury reduces materially its local corrosion in an electrolyte. The amalgamation of zinc has a twofold effect on the anodes in an electric cell. It raises the electrode hydrogen over-voltage and produces an equipotential surface over the anode that reduces local corrosion-producing couples between the zinc and metal impurities on its surface area. The amalgamation of the zinc anode is universally applied in primary-cell technology and allows use of electrolytes that would produce excessive corrosion of unamalgamated zinc. It was not until 1836 that voltaic cells became a good commercial possibility. The structural improvements of Cruikshank, Pepys, Wollaston, Hare, and others only lengthened the time before hydrogen polarization on the cathode reduced the available current flow to a negligible value. The real problem was to develop a viable chemical means for reducing or eliminating hydrogen polarization.

Chemists, although excited by Volta's epochal discovery of the means of generating a continuous flow of electricity by chemical action, recognized that it was essential to have an understanding of the chemical action involved in order to overcome the limitations of polarization.

Humphry Davy, just six months after Volta's letter to Sir Joseph Banks, reported that the oxidation of the zinc electrode in contact with the aqueous-solution spacer was responsible for generating a potential in a voltaic cell. Contrary to Volta's theory that the contact potential between dissimilar metals was responsible, he assumed a physical rather than a chemical action. Davy also found that if a strong oxidizing acid, such as nitric, was used as the electrolyte, a higher potential was obtained and polarization at the cathode was minimal. He was able to obtain an electric current flow of a much higher density, but he introduced another limitation to the life of a cell—that is, the corrosive chemical action on the electrodes independent of current flow. Thus his cell, which contained an oxidizing electrolyte that combined with the polarizing hydrogen, was an impractical solution to the problem of obtaining a stable source of electricity. Wollaston tried to minimize the chemical corrosion of the electrodes by raising them out of the electrolyte during periods of nonuse (fig. 4). For electroplating to grow

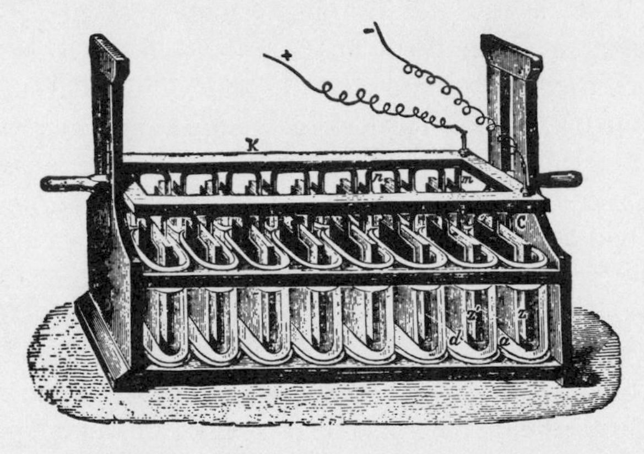

*Fig. 4. Wollaston's Trough Battery.* The plates were raised out of the solution when not in use.

into an industry, a practical high-capacity and high-energy-density voltaic cell was needed. An important advance toward this goal was contributed by William A. Grove, who proposed that the electrolyte should be two solutions (one of which was less corrosive to zinc) separated into two compartments by an ionically permeable porous cup.

Grove's battery (1838), embodying his invention, was composed of a porous cup filled with strong nitric acid in which a strip of platinum was suspended. Surrounding the cup was a weak solution of sulfuric acid around which was placed a sheet of zinc, all of the components being contained in a glass jar. Thus he reduced localized chemical attack on the zinc by the nitric acid during use or nonuse; and in the porous cup containing the nitric acid, any hydrogen ion flow to the platinum anode would be oxidized, forming water and preventing polarization, particularly at high currents. In order to make the Grove cell more economical and

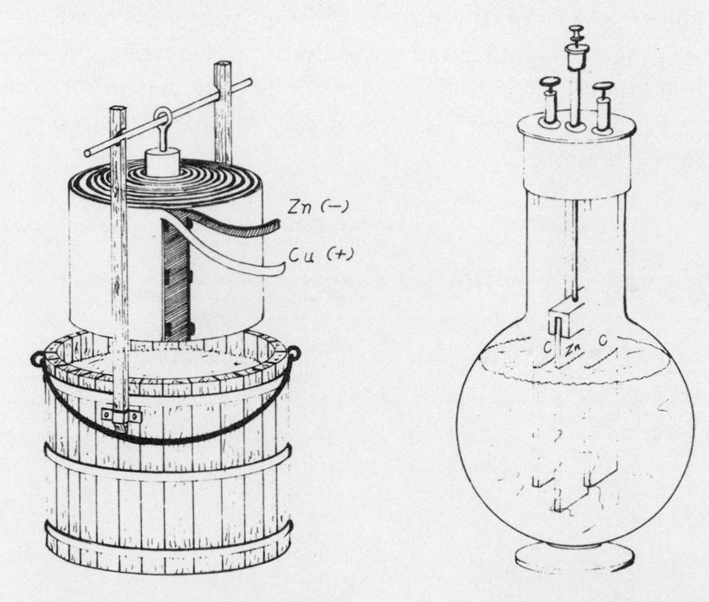


*Fig. 5. Plunge-type Hare Cell* *Fig. 6. Grenet Cell*

practical, Robert W. Bunsen in 1841 substituted a carbon rod for the platinum in the porous cup containing the nitric acid. Bunsen's cell had a voltage about twice that of the Volta cell, but it had the undesirable feature of emitting noxious nitric oxide gas caused by cathodic reduction of the nitric acid when supplying an electric current.

Johann C. Poggendorff's invention (1842), substituting chromic acid derived from a sodium bichromate sulfuric acid mixture for the nitric acid in the porous cup compartment, eliminated this problem, but at the same time his cell tended to internally polarize at high current densities, since the oxidation of the hydrogen was not sufficiently rapid. This effect was reduced by the addition of sodium chloride to the catalyte which served as a catalytic agent to effect a more rapid oxidation of hydrogen to water. The evolution of battery technology—made possible by dual-electrolyte use and separate compartments for compatible electrolytes—gave the plating industry a more practical electrochemical source of electricity.

Grenet, in 1856, produced a single-fluid, plunge-type bichromate chromic acid cell permitting the zinc anode to be raised out of the electrolyte when the cell was not in use, thus decreasing nonuse anodic reactions.

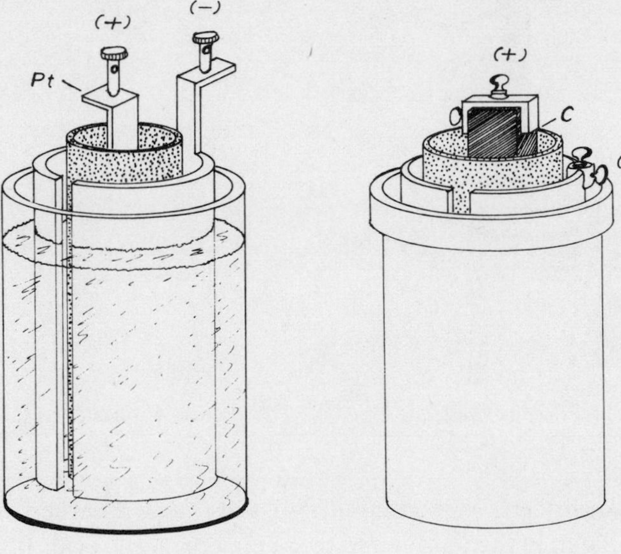


*Fig. 7. Grove Cell*

*Fig. 8. Bunsen Cell*

# CHAPTER III

*The nonpolarizing cell for telegraph industry requirements—Daniell cell—Modification of the Daniell cell, with its microporous barrier-separated electrolytes, to the gravity-positioned catalyte and anolyte in the cell of Varley.*

The rapid spread of telegraph systems, with their need for a battery capable of sustained operation over long periods, motivated the development of a battery that could be used over a long period of time without being limited by polarization or internal chemical effects.

This need was met by Professor John Frederick Daniell, whose communication "On Voltaic Combinations" to the *Philosophical Magazine* in 1836 described a nonpolarizing two-fluid cell that was destined to satisfy in principle the requirements of the telegraph industry.

The Daniell system was capable of supplying high battery capacity and continuous operation without hydrogen-cathodic polarization. It eliminated hydrogen polarization by use of a voltaic cell combination that discharged copper ions instead of hydrogen ions at the cathode ($Cu^{++} + 2e^{-} = Cu^{\circ}$).

The Daniell cell utilized a copper-strip cathode suspended in a porous cup filled with a copper sulfate solution. This cup in turn was placed in a cylindrical glass jar containing a solution of zinc sulfate and was surrounded by an amalgamated zinc cylindrical anode (fig. 9).

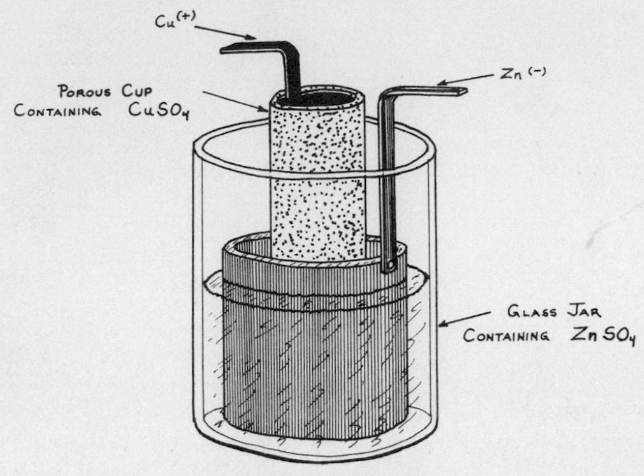


*Fig. 9. Daniell Cell*

While Daniell's porous-cup structure initially had some limitations, these were reduced by the modification from a porous-cup separation of catalyte and anolyte to a structure, credited to Cromwell F. Varley (1854), that utilized the gravity difference between zinc sulfate and copper sulfate to keep the two solutions at separate levels.

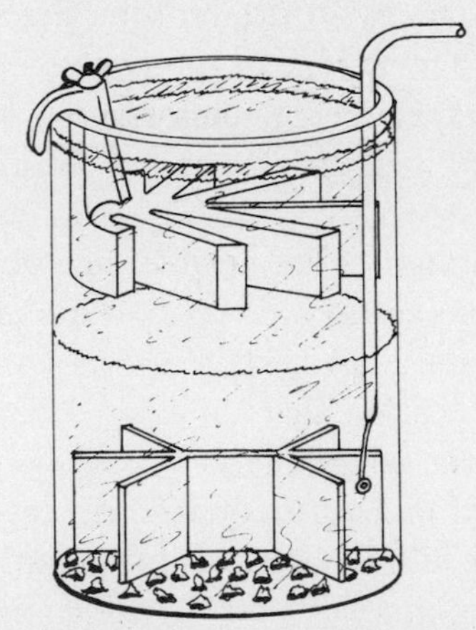

*Fig. 10. Gravity-type Daniell Cell*

The gravity cell became the standard form of the Daniell cell for most industrial applications. Fig. 10 illustrates the structure of the gravity-type Daniell cell, composed of a jar containing two layers of electrolyte. The heavier one of saturated copper sulfate occupied the bottom half of the jar and covered a copper electrode that had an insulated conductor leading to the upper surface of the jar. Carefully poured on top of the saturated copper sulfate solution was a layer of a dilute zinc sulfate solution that covered an amalgamated crow-foot-shaped zinc anode.

The gravity difference between the two solutions kept them at their levels and avoided the need of a porous cup to keep them from mixing. This allowed interionic conduction without the impedance of the small areas of the pores in the ceramic barrier. In order to maintain a saturated copper sulfate solution, an undissolved layer of copper sulfate crystals was in contact with the copper electrode.

The slight diffusion of copper sulfate into the contacting upper layer of zinc sulfate would cause some reduction effect on the zinc anode, and this was avoided by maintaining a very low continuous load current on the cell at all times.

The reactions in the Daniell cell are:

anode: $Zn \rightarrow Zn^{++} + 2e^-$

cathode: $Cu^{++} + 2e^- \rightarrow Cu^\circ$

overall cell reaction $Zn + CuSO_4 \rightarrow ZnSO_4 + Cu$

The voltage of the Daniell cell may be expressed as:

$$E = 1.1 + 0.029 \log \frac{Cu^{++}}{Zn^{++}} \text{ volts}$$

The ratio of $Cu^{++}$ to $Zn^{++}$ ions will determine the cell potential. With increased zinc ion concentration the cell voltage will be reduced, with a drop in potential from 1.17 to 1.12 volts. Amalgamation of the zinc allows a lower initial zinc ion content, and a lower pH anolyte.

The Daniell cell has a small temperature coefficient, dE/dt, between electrical energy (E) and ambient temperature (t).

# CHAPTER IV

*Recognition of railroad industry requirements—Need for adequate energy and operation under wide ambient-temperature changes for track and signal circuit operation—New type of single-electrolyte batteries by Lalande and Chaperon, using alkaline electrolytes and solid cathodic depolarizer of cupric oxide, that was capable of meeting railway requirements—Improvement of Lalande cells by Edison's bonded copper oxide cathode—Air-depolarizer alkaline cell structure having higher ampere-hour capacity and voltage than copper oxide cathode cells.*

The rapid development of the railroad industry, with its need for electrical signal systems, motivated the search for an electrochemical system that could provide a battery operable at high current density and capable of sustained operation under a wide range of ambient temperature. The large worldwide expansion of the railroad industry highlighted the need for an electric battery that had the necessary characteristics for track and signal operation.

These requirements were met by the development of the single-fluid alkaline electrolyte cell in 1882 by George Chaperon and Felix Lalande. The early Lalande cells (fig. 11) utilized as the cathodic reactant, granules of cupric oxide packed in a perforated steel container, and an amalgamated zinc anode. Both electrodes were

*Fig. 11. Lalande Alkaline Cell, $CuO$ Cathode Cell*

suspended in a glass jar containing a solution of sodium hydroxide with a thin film of mineral oil to prevent carbonation from the atmosphere. Lalande later developed an agglomerate plate of cupric oxide, which was considerably improved upon by Thomas Edison (fig. 12).

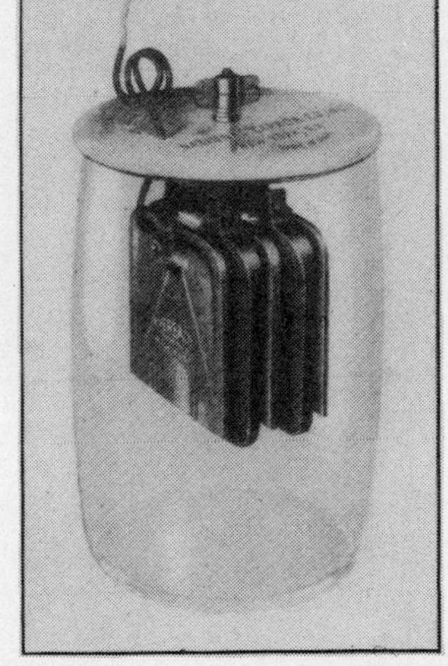

*Fig. 12. Edison-Lalande Alkaline Cell, $CuO$ Cathode Cell*

The improved Edison cathode was composed of a bonded plate structure containing cupric oxide that was pressed and bonded with tar or sodium silicate. The pressed and bonded mixture was fired at a high temperature to produce a hard and integral electrode.

A partial reduction of the cathode surface to Cu and $Cu_2O$ was effected before use to lower the internal resistance of the electrode. The high coulombic capacity obtainable by use of cupric oxide as the cathodic reactant allowed cells to be made having high ampere-hour capacity.

In order to reduce local couple action on the zinc electrode (by the dissolving of a small amount of cupric oxide into the alkaline electrolyte), a continuous load of a high-resistance circuit was kept on the cell.

The Edison-Lalande alkaline cell was produced in 1889 in a wide range of capacities. Its overall characteristics for such purposes as railroad signaling, telephone, and telegraph use were so superior to the Daniell gravity cell or the Leclanché wet cell that it displaced them. The advance effected by the use of the air depolarizer instead of the Edison bonded cupric oxide cell in turn displaced the Edison-Lalande cell for many applications.

The alkaline-electrolyte cell was improved by replacing the cupric oxide cathode by a porous carbon electrode (fig. 13) utilizing

*Fig. 13. Air-depolarizer Alkaline Cell*

atmospheric oxygen for its reducing action, thus producing a cell with many advantages. A gas had first been recognized as a cathodic reactant or depolarization element in a cell in 1839 when Grove demonstrated that oxygen was consumed at the gas/electrolyte interface of a platinum-cathode electrode and that, in combination with a similar-anode electrode to which hydrogen was applied, an electric current was produced. This was in fact the first fuel cell.

The open circuit of an air cell is 1.45 to 1.47 volts. The half-cell potential of the electrode in the 6N sodium hydroxide electrolyte is +0.13 volts for the cathode and −1.32 volts for the amalgamated zinc anode. The closed circuit voltage depends upon the plate area and the load currents. The load voltage is about twice that obtained with the cupric oxide cathode cells.

The occurrence of air depolarization in the early wet cells (zinc/ammonium chloride [aqueous]/manganese dioxide) was also noted by Leclanché. He observed that by only partially immersing the carbon manganese dioxide electrode mix in the electrolyte, as much as 50 percent of the cell output at low current densities could be ascribed to the oxygen in air depolarization of the cathode. The complete use of air as the cathodic reactant or depolarizer was described by L. Maiche in his British patent (no. 1940, Oct. 24, 1882).

Junger, in his U.S. patent (no. 913,340, Feb. 23, 1909) on electrodes for gas elements, described the enhancement of the carbon cathode activity by incorporating catalysts such as finely divided silver, palladium, platinum or activated carbon; he also suggested an electrode structure with a capillary portion to retain the electrolyte with a coarse hole structure (macropore), permitting air access to the interior of the electrode and increasing the surface of contact of the three phases: air, electrolyte, and conductor. These basic stipulations have been of importance in modern studies in primary air and fuel cells.

Because of the shortage of battery-grade manganese dioxide ore, there was a considerable increase in the development of air-depolarizer primary cells during World War I. In U.S. patent (no. 1,678,405, July 24, 1928), R. Oppenheim described the production of air-depolarizer cells by wetproofing the porous carbon electrode. They made it impervious to aqueous electrolytes without impairing permeability to air and electrical contact to the active cathode surfaces.

The early cells utilizing air-depolarizer cathodes with zinc anodes and ammonium chloride electrolyte were made in single-battery sizes up to 1000 ampere-hour capacity. The use of alkaline electrolytes, with their better electric conductivity, smaller temperature conductivity coefficients, and lower freezing points, was described (in U.S. patent no. 524,229, August 1897), by Walker, Wilkins and Lones. When embodied in later air-depolarizer cells it superseded for most purposes the earlier ammonium chloride electrolyte.

The use of activated carbon in the carbon cathode permitted a cathode structure not requring expensive catalytic agents. In some applications higher current densities—ten times that obtainable with air as a cathodic reactant—have been reported when pure oxygen has been discharged into the active porous carbon-cathode structure.

One requirement for the porous carbon cathodes is that since the oxygen is diffused to the interfaces with the electrolyte, it is necessary that the porous carbon be resistant to flooding of its active porous structure by the electrolyte. Various wetproofing means have been used, such as paraffin film, petroleum oils, and silicone or other hydrophylic coatings.

An increase in the electrolyte capacity and its maintenance of conductivity is provided by the addition of calcium hydroxide granules to the alkali electrolyte. The operation of the zinc-anode alkaline cell involves production of a zincate as an anode product, such as sodium zincate when sodium hydroxide is the electrolyte. The sodium zincate dissolving in the alkali hydroxide decreases the conductivity of the electrolyte, thus lowering the cell efficiency by increasing the electrolyte voltage drop on load. The calcium hydroxide, when added, combines with the zincate and precipitates it as an insoluble calcium zincate and releases the alkali for maintenance of its concentration. This could be expressed as follows:

$$2NaZn(OH)_4 + Ca(OH)_2 + 2H_2O \rightarrow CaZn_2O_3, 5H_2O + 4NaOH$$

The alkaline-electrolyte air-cathodic reactant cells vary in sizes from 300 ampere-hours to 1500 ampere-hours in single units.

# CHAPTER V

*Development by Leclanché of cells having single-fluid electrolytes of low anodic shelf oxidation with solid-oxide cathodic reactants of manganese dioxide—Applications of liquid-electrolyte Leclanché cells—Modification of Leclanché system of zinc-ammonium chloride-manganese dioxide-carbon by Gassner, resulting in a dry-cell structure utilizing the zinc anode as a container of cell components—Later structural and material improvements—Application of dry cell to telephone and automobile industries.*

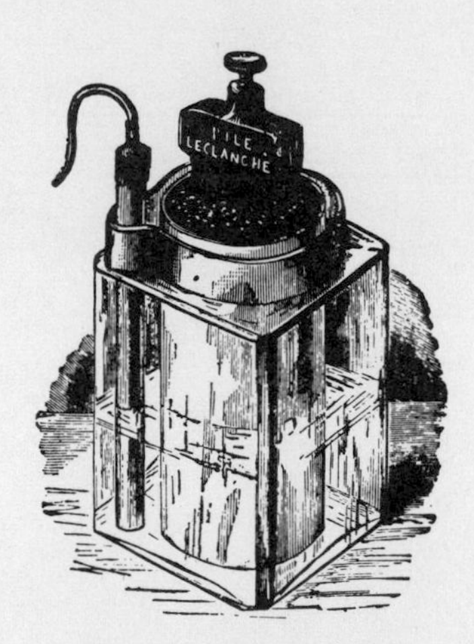


*Fig. 14. Leclanché Cell*

The properties of metallic oxides, such as manganese dioxide, as cell cathodic reactants had been recognized by Volta and Zamboni.

Michael Faraday wrote in his *Experimental Researches in Electricity* (Vol. I, [1849]) that the electromotive force generated in a voltaic cell may be combined potentials generated at both the cathode and anode. He referred to the studies of Bequerel, De La Rive, and Muncke, as well as his own, on cell components. He listed the following order of materials which he found would contribute cathodic potential in acidic electrolytes. These were listed in an order relative to metallic anodes.

Peroxide of lead
Peroxide of manganese
Oxide of iron
Plumbago
Rhodium
Platinum
Antimony
Copper
Zinc

Plumbago was listed as the neutral substance. The materials above plumbago had an increasing potential difference to those below. He noted that M. Muncke, in his studies of cathodic materials, observed the ability of lead peroxide to react potentially like manganese dioxide.

The industrial need was for a single non-corrosive fluid electrolyte cell capable of providing a good potential and long shelf life. These characteristics were provided by Leclanché, who in his system used an amalgamated zinc anode, an electrolyte of ammonium chloride, and a cathode reactant of manganese dioxide granules mixed with carbon.

### *Single-Fluid Solid-Depolarizer Cells*

The first practical single-fluid cell with a solid cathodic reactant was built by Leclanché in 1866. He found that manganese dioxide, when mixed with carbon, possessed a good electrical conductivity, capable of acting as a depolarizer or cathodic reactant by being reducible to a lower oxide on passage of electric currents. He mixed ground manganese dioxide with an equal quantity of retort carbon and pressed this mixture and a carbon rectangular rod into a porous cup. This in turn was placed in a square glass jar containing a solution of ammonium chloride and an amalgamated zinc anode (fig. 14). The depolarizing cathode mix was contacted by a pressed carbon plate. Leclanché's cells had a potential of 1.5 volts, or 1.382 times that of the Daniell cell.

Later, in 1876, Leclanché described a modification of his earlier cell in which a mix of manganese dioxide, carbon and a resin binder was pressed into an aggregate block, thus dispensing with the porous-cup cathode container. He used a mixture comprising 40 percent manganese dioxide, 55 percent carbon, 5 percent gum resin; and to maintain good porosity for more uniform contact in depth with the electrolyte, he added 3-4 percent of potassium bisulfate. Hydraulic pressure and a temperature of 100°C. were used to process the cathode element into a strong plate or cylinder. He also found that allowing free space about the electrolyte afforded depolarization by the oxygen in the air and increased the cell capacity.

The Leclanché cell found widespread application. However, it still lacked one feature that would greatly extend the application of the system, namely, a more portable, unspillable structure. A number of investigators experimented with various means for immobilizing the ammonium chloride electrolyte utilizing such materials as plaster of Paris, sawdust, chalk, zinc oxide, and gels made with starch or sodium silicate.

*The Leclanché-system Dry Cell*

The rapid expansion of the Leclanché system of zinc, ammonium chloride, and manganese dioxide/carbon was due to the development of the system into a "dry cell." Gassner is credited with developing a dry-cell structure utilizing a zinc container as the anode and an immobilized electrolyte; the latter consisted of a paste made with plaster of Paris and ammonium chloride and was separated from the anode by a cloth bag-enclosed cathode of carbon and a depolarizing oxide. Pitch was poured on top and around a protruding carbon rod that contacted the manganese dioxide/carbon cathode. In a later form (fig. 15) the electrolyte was added to the

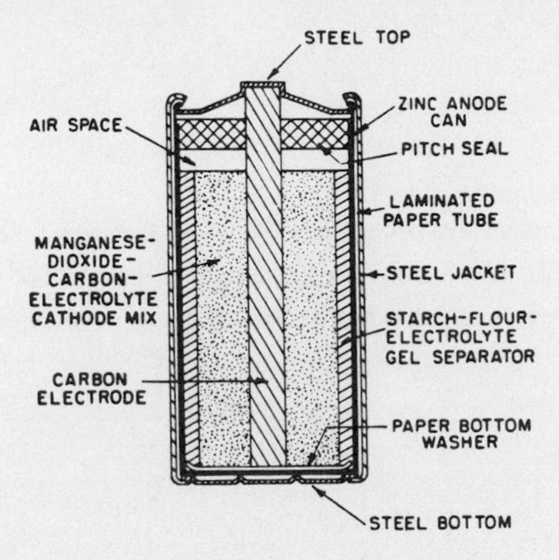


*Fig. 15. Leclanché-system Dry Cell*

carbon/manganese dioxide mixture and pressure formed into a bobbin which was separated from the zinc-can anode by an absorbent cellulose spacer liner and a starch gel containing zinc chloride and some mercuric chloride for amalgamation of the zinc anode container. The carbon was wax impregnated to prevent external electrolyte creepage, but it remained porous enough to allow gas to escape to a perforated brass contact cap. In present-day cells, for better protection against the effects of leaking gel electrolyte when the zinc is perforated at end of life, a steel jacket is formed around the zinc can and insulated by a grommet from the top contactor.

Few developments in the history of science have had so universal an application as the Leclanché-system battery. Its essential chemistry has not changed since its conception by Leclanché in 1876, although it has evolved into a dry-cell structure.

A characteristic of the Leclanché-system dry cell is that the ampere-hour capacity of the Leclanché-system dry cell is determined by such factors as rate of discharge, length of discharge time, and ambient temperature.

The effects of these in a #6 cell can be noted in the following:

| *Discharge amperes* | *Hours of discharge per day (24)* | *Capacity in ampere-hours* |
|---|---|---|
| 0.10 | 4 | 40 |
| 0.10 | 8 | 36 |
| 0.10 | 24 | 28 |
| 1.25 | 24 | 7.3 |
| 2.50 | 24 | 4.6 |
| 5.0 | 24 | 23 |

The improvement in performance of the #6 size of Leclanché-system cells over the years in telephone circuitry of low current discharges can be seen from the following:

| *Year* | *Days of effective use* |
|---|---|
| 1910 | 165 |
| 1920 | 230 |
| 1930 | 360 |
| 1935 | 450 |

The progressive improvement of the cell since 1910 is related to manganese dioxide of improved purity and larger percentages of zinc chloride in the electrolyte.

An important early industrial application of the #6 cell was for ignition in the internal combustion engine. This application involved production of a self-induction high potential and an arc between two electrodes in the combustion chamber; when mechanically separated at a proper time interval the latter selectively open and close the highly inductive low-resistance circuit.

This method was superseded by the use of a high-voltage inductance coil that, in a secondary winding, developed a high potential adequate to discharge a spark through the gap in a spark plug. The primary winding was continuously interrupted by an electromagnetically operated interrupter, and its secondary winding supplied the high potential to a distributor that selectively applied the high voltage to the proper spark plug in synchronism with the position of the cylinders.

The recognition that cells with higher potentials would require the use of anodic material having the highest oxidation potential led to the application of the alkali and alkaline earth metals as anodes. The values shown below are comparative to the oxidation of hydrogen (at one atmosphere pressure) to a hydrogen ion (in a one-molar acid solution of the metal at 25 °C.).

$$Li \rightarrow Li^{+} + e = 3.045 \text{ volts}$$
$$K \rightarrow K^{+} + e = 2.925 \text{ volts}$$
$$Na \rightarrow Na^{+} + e = 2.714 \text{ volts}$$
$$Mg \rightarrow Mg^{++} + 2e = 2.363 \text{ volts}$$
$$Zn \rightarrow Zn^{++} + 2e = 0.7628 \text{ volt (Basic 1.32 volts)}$$
$$Cd \rightarrow Cd^{++} + 2e = 0.4029 \text{ volts}$$
$$H_2 \rightarrow 2H^{+} + e = 0.000 \text{ volt (Basic 0.83682 volts)}$$

Early attempts to use anodic materials of the highest potential failed because of chemical reaction with the inorganic electrolytes. Solid sodium and potassium amalgams were reported in use as electrode materials in 1847 but were not found practical. Cells using magnesium as the anode in the same type of structure as the Leclanché dry cell (fig. 6) with manganese dioxide/carbon cathodes have potentials in the order of 1.9 volts and are used to a limited extent in military applications.

The development of the magnesium cell required a study of compatible electrolytes, magnesium alloys, and inhibiting agents. The pH of the electrolyte is an important factor in controlling the rate of shelf corrosion of the magnesium anode. Cells, made by the Dow Company, using alloys of magnesium, such as their alloy AZ10A (1 percent aluminum, 0.5 percent zinc and 0.2 percent calcium), have shown improved shelf life and reduction of initial time required for a cell to reach its operating potential.

The basic problem of chemically more active anodes, when used with aqueous electrolytes, is the anode efficiency. This becomes a question of the ratio of the number of faradays obtained by ionization of the anode to the number of equivalents dissolved by local chemical action. As much as 40 percent of the magnesium anode can be consumed by local corrosion action during discharge, as compared to 5 percent or less with zinc anodes. With magnesium, an ionic gradient can develop on open circuit by the combined magnesium oxide layer and the hydrogen collected in its pores. The alloys tend to reduce the time delay effect from seconds to a lesser amount, depending on the area of the magnesium electrode, its prior time of discharge, and pH change of electrolyte.

The use of magnesium perchlorate electrolyte with the low-percent-aluminum and -zinc alloys of magnesium have been helpful in making the magnesium anode more viable. Its performance is better than with electrolytes of magnesium bromide and chromate inhibitors. Although the magnesium cell has a voltage of 1.90 and although there has been some improvement with the use of alloys, the cell appears to have inherent time-delay limitations which have restricted its use for a number of applications.

Aluminum as an anode has the same problem to a degree, and less coulombic efficiency. Aluminum anodes with electrolytes of manganous or aluminum chloride with a chromate inhibitor have been used, the cells having a potential of 1.5 volts. The overall performance does not surpass that of the zinc anode cell when the latter is made with electrolytic-grade manganese dioxide and alkaline electrolytes.

With some aerospace and marine applications requiring voltages higher than those obtainable from zinc or magnesium anode cells, and of higher energy density, progress in this direction has been made by the development of the sealed lithium anode cell having a non-hydrogen-polarizing organic electrolyte and a carbon fluoride or sulfur dioxide cathodic reactant.

The comparative values of coulombic capacity of present-type primary cell systems are as follows:

| *Cell system* | *Voltage E* | *Capacity* watt-hours per pound | *Capacity* watt-hours per kilogram |
|---|---|---|---|
| $Zn/NH_4Cl$ aq./$MnO_2$ + carbon | 1.5 | 30 | 66.67 |
| Zn/KOH aq./$MnO_2$ + graphite | 1.5 | 35 | 77.70 |
| Zn/KOH aq./HgO + graphite | 1.345 | 50 | 110.0 |
| Zn/KOH aq./$Ag_2O$ | 1.6 | 60 | 133.2 |
| $L_1$/PC* + $LiClO_4$/$CF_n$ | 3.3 | 200 | 440 |

*PC = Propylene carbonate as solvent.

# CHAPTER VI

*World War II requirements: transreceiver operation; detection equipment; remote control equipment; telemetric systems—Sealed alkaline systems: zinc mercuric oxide sealed alkaline cell of R.M. system—Postwar application: miniaturization for bioelectric application; zinc/silver oxide alkaline sealed cells for electric watch and digital calculators.*

## *Military Requirements and the Development of Sealed Alkaline Cells*

The advent of World War II and military requirements for a dry cell having special characteristics motivated the development of the zinc/mercuric oxide sealed alkaline cell. Determination of the type of cell suitable for a given application was no longer a matter of cell geometry, as it had been in the past, but one of an electrochemical system that would operate most effectively for a particular use. The size of the cell was related more to the total capacity or energy developed for the load range, and not so limited in respect to the rate of discharge. This was highlighted in 1941 by the advent of large-scale use of transreceivers and miniaturized equipment where more exacting requirements were encountered. There were three requirements; namely, maximum electrical capacity per unit volume, maximum storage life under varying ambient temperatures, and as constant a discharge voltage as possible.

One of the problems related to logistics met in military applications, particularly where the fields of use were thousands of miles away from the manufacturing source, was that the ambient temperatures and humidity were high enough to affect adversely the capacity of the Leclanché-system cell. In World War II, the area of the South Pacific was particularly demanding in its requirements of batteries.

In 1942 the development of small sealed alkaline cells was commenced at the Ruben Laboratories, utilizing the experience obtained many years earlier in their development of the hermetically sealed C-bias voltaic cell and the dry electrolytic capacitor. The anode of the bias cell (fig. 16) was a cadmium con-

*Fig. 16. Hermetically-sealed Grid-bias Cell*

tainer separated by a grommet from, and crimped to, a cathode composed of a sintered disc of vanadium pentoxide serving as the top contact. The gel electrolyte which filled the container was composed of ammonium glycoborate, and the cell was hermetically sealed. Since there was minimal chemical reaction between the cell elements on shelf or during operation, the cell had, for practical applications, a life of many years. The first sealed alkaline cell structure utilized a wound structure similar to the Ruben capacitor structure (shown in fig. 17). The wide-area rolled anode was closely spaced to its rolled cathode for low internal resistance. The dry electrolytic capacitor had two aluminum foil strips that were separated by a 0.25-millimeter paper spacer impregnated with ammonium glycoborate and sealed in an aluminum container. The cathode in a cell of the wound anode and cathode (fig. 18) structure was a 2-millimeter sandblasted steel strip coated with a layer of cupric oxide and graphite powders and bonded with a 5 percent

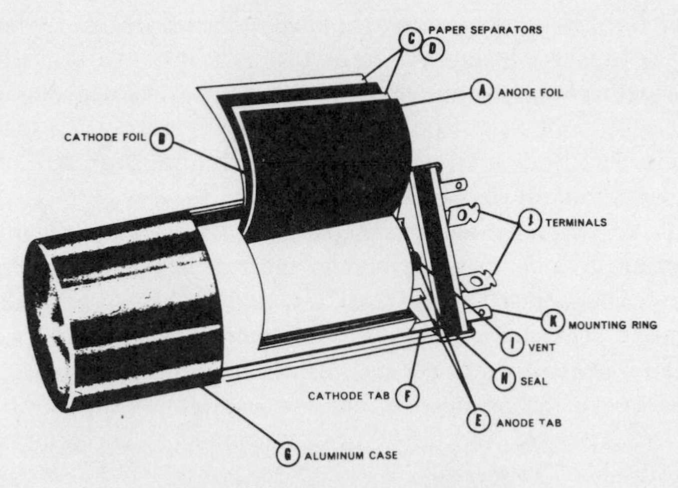


*Fig. 17. Dry Electrolytic Capacitor*

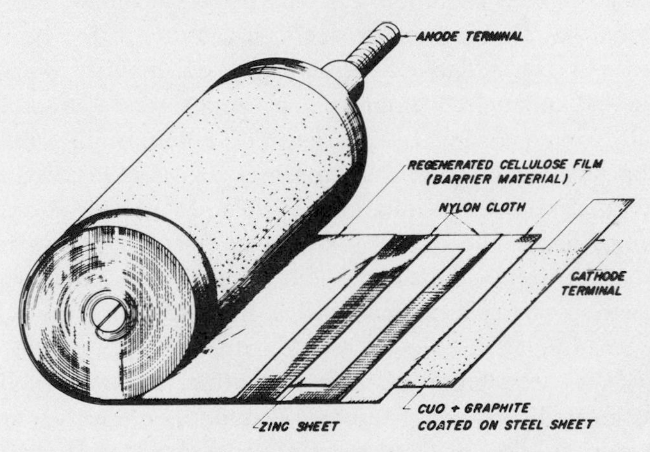


*Fig. 18. Wound-Anode Zinc/Mercuric Oxide Sealed Alkaline Cell (R.M.)*

polystyrene solution. This coated strip, after baking at 100°C., was mechanically rolled at high pressure, producing a low-resistance flexible cathode layer. The spacers were a combination of an ionically conductive cellophane layer, as a barrier in contact with the oxide to prevent migration of cathode particles, and an absorbent nylon gauze element in contact with the zinc. A dusting of powdered mercuric oxide was applied to the anode before winding, so as to cause amalgamation of the surface of the zinc when in contact with the alkaline electrolyte. In order to reduce solubility of the cupric oxide in the potassium hydroxide electrolyte, a solidified high-concentration electrolyte was used and impregnated into the rolled unit at the electrolyte melting point.

After impregnation and electrolyte solidification, the unit was fastened to a composite neoprene and bakelite insulating top that was crimp-sealed to the metal can and had a porous plug for venting. The cupric oxide cell coulombic capacity and its initial voltage of 1.06 were not adequate, and other oxide coatings such as mercuric oxide, manganese dioxide, cadmium oxide, and silver oxide processed in the equivalent manner were tested comparatively. The mercuric oxide (1.345 volts) produced the best overall results. Since the main object was to provide a sealed alkaline cell with maximum capacity per unit volume, the design was changed to one using a steel-can cathode in the bottom of which was consolidated a densely pressed mixture of mercuric oxide and micronized graphite. The anode was composed of a spirally wound corrugated zinc strip with an intervening absorbent cotton paper spacer. The winding was in an offset manner, so that the corrugated zinc extended at one end and the paper at the other. Corrugation of the anode was found essential to provide the necessary space for the zinc oxide produced at the anode when the cell was discharged, to maintain a free electrolyte path. After impregnation of the wound anode-absorbent spacer unit in the solution of zincated potassium hydroxide, the wet anode was amalgamated by applying a measured quantity of mercury (10% of zinc weight) to the exposed turns of the corrugated zinc after immersion of the unit in the electrolyte at 50°C., at which temperature the mercury diffused rapidly into the spirally rolled anode. This provided an amalgamated top anode contact to the external terminal and an interface junction without local galvanic action. The barrier disc interposed between the extending paper end of the anode and the mercuric oxide cathode was a microporous dialysis paper disc; other barriers such as sintered magnesium oxide discs were also used. The potassium hydroxide electrolyte was zincated for minimum zinc corrosion on shelf by dissolving 16 grams zinc oxide in 100 milliliters of a 40 percent solution of KOH at a temperature of 100°C.

In order to obtain maximum anode area, another type of spirally rolled anode was experimentally tested in which zinc was sprayed by a "Schori" metal spray gun on one side of a cellophane strip 0.009 inch thick, producing a porous layer deposit of sintered zinc powder. The zinc coated spacer was then corrugated, wound in combination with a Dexter paper layer in contact with the zinc, and amalgamated in the same manner as the solid zinc-strip anode. However, while large anode areas were obtained, they did not result in uniformly suitable cells because of impurities derived from the heating gases used to volatize the zinc in the spraying equipment. In later cells, the use of pressed-powdered-zinc amalgam served the same purpose without the impurity factor experienced with the spray method.

A most important development in sealed alkaline cells occurred at the Ruben Laboratories with the conception of the coulombmetrically balanced cell having equivalent anode and cathode capacities. Since the hydrogen deposition potential is 0.898 volt and the container overvoltage is 0.15 volt, the sum of both is less than the half-cell potential of zinc in the electrolyte, or 1.332 volts. Any residual unoxidized zinc would cause hydrogen generation at the cathode on continued application of load after the oxygen content of the depolarizer had been exhausted. A balanced cell also basically required that the anode top contact be composed of a non-consumable but amalgamatable material such as copper or tinned steel. In order to compensate for errors in the required measured weights of the anode and cathode materials, an excess amount of cathode was used; in most cases the addition of a small percentage of synthetic manganese dioxide to the more expensive mercuric oxide was adopted for lower costs.

*Commercial Cell Structures*

On pages 56 and 57 are illustrated the three types of sealed-cell construction embodying the basic factors found necessary for producing sealed alkaline cells. While the designation and operational data are related to cells having mercuric oxide as their depolarizer or cathodic reactant, the same structures apply to cells wherein manganese dioxide or silver monovalent or divalent oxide is used as the cathodic reactant. The silver oxide cells require a noncellulosic and more oxidation-resistant barrier disc in contact with the oxide than do the cells with mercuric or manganese dioxide.

Figures 19, 20, and 21 illustrate the geometry of the present "RM"*-type cells of sealed zinc/mercuric oxide/alkaline cells.

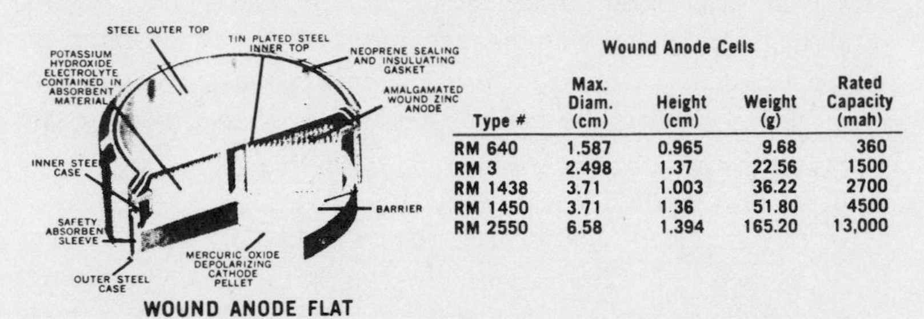


**Wound Anode Cells**

| Type # | Max. Diam. (cm) | Height (cm) | Weight (g) | Rated Capacity (mah) |
|---|---|---|---|---|
| RM 640 | 1.587 | 0.965 | 9.68 | 360 |
| RM 3 | 2.498 | 1.37 | 22.56 | 1500 |
| RM 1438 | 3.71 | 1.003 | 36.22 | 2700 |
| RM 1450 | 3.71 | 1.36 | 51.80 | 4500 |
| RM 2550 | 6.58 | 1.394 | 165.20 | 13,000 |

*Fig. 19. Wound-Anode/Cathode of Early Sealed Alkaline Cell (R.M.)*

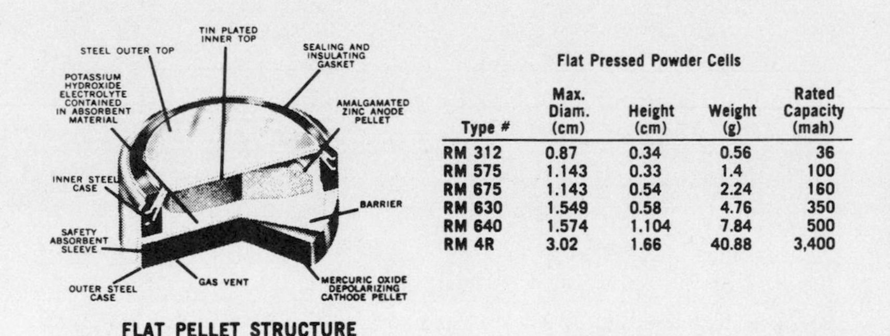


**Flat Pressed Powder Cells**

| Type # | Max. Diam. (cm) | Height (cm) | Weight (g) | Rated Capacity (mah) |
|---|---|---|---|---|
| RM 312 | 0.87 | 0.34 | 0.56 | 36 |
| RM 575 | 1.143 | 0.33 | 1.4 | 100 |
| RM 675 | 1.143 | 0.54 | 2.24 | 160 |
| RM 630 | 1.549 | 0.58 | 4.76 | 350 |
| RM 640 | 1.574 | 1.104 | 7.84 | 500 |
| RM 4R | 3.02 | 1.66 | 40.88 | 3,400 |

*Fig. 20. Pressed-powder Zinc/Mercuric Oxide Sealed Alkaline Cell (R.M.)*

---

*"R.M." is the designation applied by the U.S. Signal Corps during World War II as an acronym based on the names "Ruben" and "Mallory," the inventor and the principal manufacturer, respectively, of the sealed Zn/HgO cell.

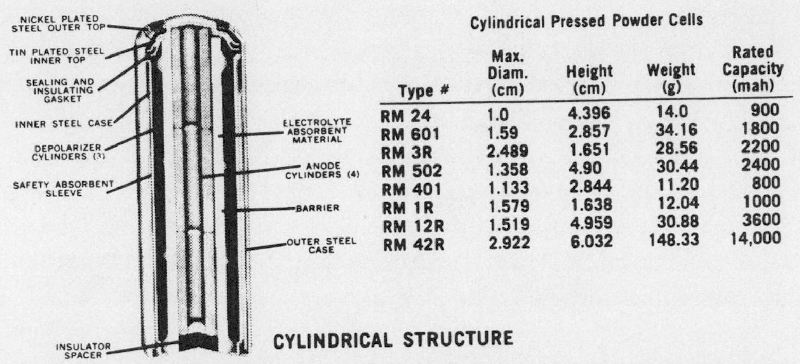


**Cylindrical Pressed Powder Cells**

| Type # | Max. Diam. (cm) | Height (cm) | Weight (g) | Rated Capacity (mah) |
|---|---|---|---|---|
| RM 24 | 1.0 | 4.396 | 14.0 | 900 |
| RM 601 | 1.59 | 2.857 | 34.16 | 1800 |
| RM 3R | 2.489 | 1.651 | 28.56 | 2200 |
| RM 502 | 1.358 | 4.90 | 30.44 | 2400 |
| RM 401 | 1.133 | 2.844 | 11.20 | 800 |
| RM 1R | 1.579 | 1.638 | 12.04 | 1000 |
| RM 12R | 1.519 | 4.959 | 30.88 | 3600 |
| RM 42R | 2.922 | 6.032 | 148.33 | 14,000 |

*Fig. 21. Cylindrical Zinc/Mercuric Oxide Sealed Alkaline Cell (R.M.)*

They are, respectively, the wound-anode flat type, the flat-type cell utilizing pressed-zinc-amalgam pellet as the anode, and cylindrical-structure cells with pressed-zinc-amalgam cylinder anodes and pressed mercuric oxide/graphite cathodic-reactant cathodes.

With larger cells, an added safety precaution against possible gas-producing impurities in the cathode material is provided by using a pressure-vented structure with an outer steel container separated by a cardboard tube that absorbs any electrolyte that may be vented. This extra container is not required on the smaller button-type cells as used for portable transistorized circuits.

Natural manganese ore, because of its variable ferrous content, was found to produce on the zinc anode dendritic growths that extended through the barriers and caused interelectrode leakage paths. The advent of electrolytically produced manganese dioxide in the postwar period made more practical sealed alkaline cells employing manganese dioxide and graphite cathodes. While having neither the equivalent coulombic capacity in a given cell volume, nor the maintenance of load voltage during discharge of the mercuric oxide cathode, its lower cost rendered it economically preferable for numerous industrial applications. Its commercial application has been rapidly increased with production on a high-volume scale.

The operation of the cell utilizing mercuric oxide as the cathode should be stated as:

$$Zn \rightarrow Zn^{++} + 2e^-$$
$$Zn^{++} + 4OH^- \rightarrow Zn(OH)_4^{--}$$
$$Zn(OH)_4^{--} + 2K^+ \rightarrow KZn(OH)_3 + KOH$$
$$KZn(OH)_3 \rightarrow ZnO + KOH + H_2O$$
$$HgO + H_2O \rightarrow Hg(OH)_2 \rightarrow Hg^{++} + 2OH^-$$
$$Hg^{++} + 2e^- \rightarrow Hg, 2OH^- + 2H^+ \rightarrow 2H_2O$$

Overall chemical reaction for producing 2 faradays per gram mole of anodic zinc and gram mole of cathodic mercuric oxide is

$$Zn + HgO \rightarrow ZnO + Hg$$

Since the basic reactions are the oxidation of the zinc, $Zn \rightarrow Zn^{++} + 2e^-$, and the reduction of the mercuric oxide, $Hg^{++} + 2e^- \rightarrow Hg$ at the cathode, and since water appears at both electrodes, there is no significant change in electrolyte composition.

The anode of the wound-anode flat type structure is composed of a corrugated zinc strip with a paper absorbent wound in an offset manner so that it protrudes at one end and the zinc protrudes at the other end. The zinc is amalgamated with mercury (10 percent by weight) after the paper spacer is impregnated with the electrolyte, which impregnation also causes the paper to swell and produce a positive contact pressure to the electrode. The other necessary details of construction are apparent in the illustrations.

In the pressed-powder cells, the zinc powder is preamalgamated prior to pressing into shape, and its porosity allows electrolyte impregnation with oxidation in depth when current is discharged. A double-can structure is used in the larger sized cells, the space between the inner and outer containers providing passage for any gas generated either by a cell not properly balanced in relation to the stoichiometric requirements of anode to cathode weights or by the presence of impurities in the anode material. The construction is such that if excessive gas pressures are experienced, the compression of the upper part of the grommet by internal pressure allows the gas to escape into the space between the two cans. A paper tube surrounds the inner can so that any liquid carried by discharging gas will be absorbed, maintaining a leak-resistant structure. Release of excessive gas pressure automatically reseals the cell.

*Cell Discharge Characteristics*

The general characteristic of the Zn/HgO system is noted in fig. 22. This figure illustrates an equilibrium period ($T_2$) representing practically 97 percent of the cell life when the current drain is relatively low; e.g., for an RM-1 cell at 1 milliampere drain, voltage is constant (at 98 percent of no-load potential) within less than 1/2 percent during this period. At higher level drains, such as 20 milliamperes, the potential will vary by +2 percent from 90 percent of no-load level, and the equilibrium period is approximately 67 percent of cell life. Larger cells, or groups of smaller cells in parallel, would show longer equilibrium-period characteristics for equivalent rates of drain. Since the current density per unit of electrode area determines the equilibrium period, it is obvious that in a proper cell design the voltage may be as constant as desired (for any predetermined equilibrium period) to meet the demands of any particular circuit. The electrical and mechanical specifications for the

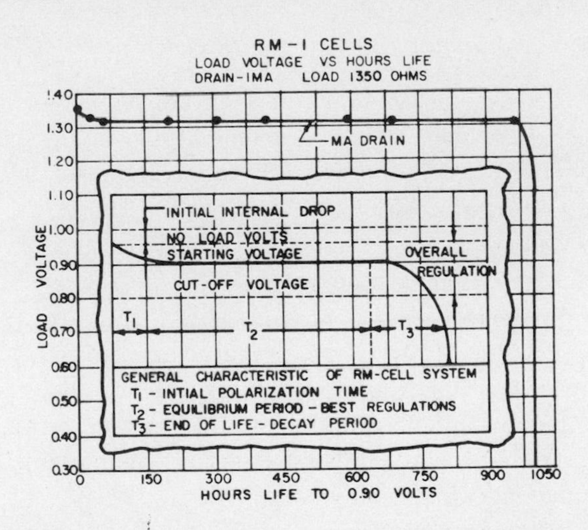


*Fig. 22. Graph of General Characteristics of Zn/HgO System*

RM-1 cell are 1.580 centimeters diameter, 1.638 centimeters height, 1.960 cubic centimeters volume, 12.191 grams weight, and 1000 milliampere-hour capacity.

Maintenance of voltage during discharge was one of the important characteristics of the sealed zinc mercuric oxide cell that made it of special interest for transreceiver use in military applications and for use in electric watch and implantable cardiac pacemakers during the postwar period. Fig. 23 illustrates the relative discharge characteristics, in relation to time and operating capacity, of a Leclanché-system dry cell and a sealed zinc/mercuric oxide alkaline cell.

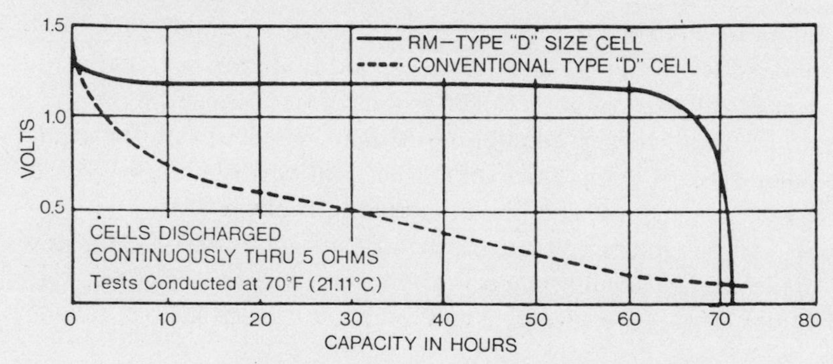


*Fig. 23. Graph of Comparative Discharge of Zn/HgO*

One of the basic factors that causes this difference is that in the Leclanché system the cathodic reactant $MnO_2$ changes from fair semiconductivity to the practically nonconductive $Mn_2O_3$. In operation, we have 2 $MnO_2 \rightarrow Mn_2O_3$, whereas in the mercuric oxide cell we have an increase in conductivity because of the direct reduction of the semiconductor HgO to metallic mercury Hg. Increase of electrolyte pH at the $MnO_2$ cathode and lower reduction potential of $Mn_2O_3$ on discharge are also responsible for decreasing cell voltage during discharge. Since the number of electrons in one coulomb is $6.28 \times 10^{18}$, a rapid reduction without an appreciable ionic gradient is necessary for maintaining the cathode potential under load conditions.

One of the most gratifying applications of electric battery

technology has been the use of the sealed zinc/mercuric oxide cell for bioelectrical devices, particularly for implanted cardiac pacemaker units. The basic requirement of a cell for bioelectric use has been chemical and electrochemical stability.

*Fig. 24. Implanted Cardiac Pacemaker with Zn/HgO Sealed Alkaline Cell*

The first cardiac pacemaker units utilized the standard 1-ampere-hour commercial cell structure that had been designed to perform efficiently at relatively high current densities; however, the cardiac pacemaker requires operation at very low current densities for a long time. To obtain maximum capacity it is necessary to avoid internal bridging paths which could be produced by migration of minute amounts of mercury or zinc oxide under ionic, electrophoretic, or mechanical forces through the spacers. This is a problem not encountered with other applications requiring discharges of high current for shorter periods of time.

Ruben reported (in the *Transactions of the New York Academy of Science* 161 [October 1969]: 617-634) that the relative discharge characteristics and capacity of the cardiac pacemaker production cells had depended in part on mechanical retention in the cathode of any mobile mercury cathodically produced in the mixture of mercuric oxide and micronized graphite. The Zn/HgO cell discharge characteristics have been duplicated with cathodes composed of finely divided silver instead of graphite as the electrical contact to the mercuric oxide. This has eliminated any electrophoretically induced leakage path problems with finely divided mercury, since the silver combines immediately with the reduced mercury to form an immobile solid amalgam. The addition of finely divided silver to the mercuric oxides also allows recharging of the cell in other applications where a rechargeable cell can be used to advantage. An automatic charger cutoff voltage is produced when the mercury has been reoxidized to mercuric oxide. Further oxidation of the silver component raises the cell potential from 1.345 to 1.56 volts with $Ag_2O$ formation, or, if continued, to 1.8 volts and AgO formation. The oxidation of the silver component is only superficial and only sufficient to raise cell potential to a relay cutoff value of the charger. A short discharge quickly reduces the cell overcharge potential to its Zn/HgO value of 1.345 volts.

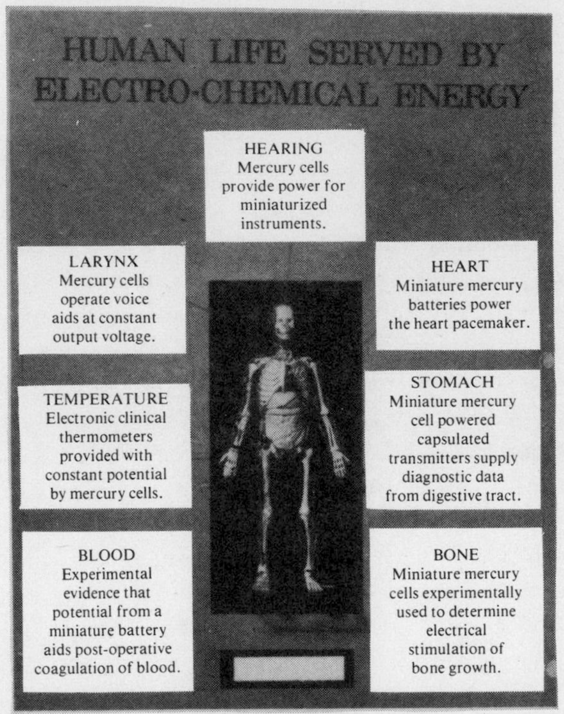


*Fig. 25. Illustration of Bioelectric Application of Zn/HgO Sealed Alkaline Cell*

Extreme miniaturization of electronic components, coupled with dependable performance, has steadily gained importance in electronic applications. The development of miniaturized electric cells with adequate coulombic capacity has been enhanced by the motivating influence of such projects as telemetric devices in the upper atmosphere and lightweight instruments for detection of radiation, as well as a growing realization of the necessity for continuous advance in all phases of modern battery technology.

The application of miniaturized button-type cells to compact radios, transreceiver hearing aids, implantable cardiac electric pacemakers, and other applications followed the development of the sealed zinc/mercuric alkaline oxide cells. For some sealed alkaline cell applications, where a potential higher than 1.345 volts is necessary, the mercuric oxide is replaced by silver oxide which in the monovalent form provides a potential of 1.50 volts and in the divalent form, 1.82 volts. The silver oxide cells have been particularly useful in the operation of light emitting diodes (L.E.D.) which require a higher voltage than the 1.35 volts obtained with mercuric oxide cells. Silver oxide also has the advantage of not requiring the addition of an electron-conductive component to the cathodic reactant so that the maximum depolarizer capacity can be obtained from a given weight of silver oxide. The cathode resistivity continuously decreases with reduction of the silver oxide to silver, thus enabling the maintenance of a high output voltage during its useful cell life.

### *Nonalkaline Mercuric Compound Cells*

*Zinc Mercuric Dioxysulfate Cell.* Another system has been developed for lower-discharge current requirements which has some of the desirable flat-voltage-discharge characteristics of the zinc/mercuric oxide alkaline cell, while allowing the use of the less expensive conventional dry cell construction. In the past, attempts had been made to use mercury compounds in the cathode of a nonalkaline dry cell, but these failed because of limitations inherent in the materials used. In a cell developed at the Ruben Laboratories, which might be termed the mercury/zinc/carbon cell, the cathodic reactant is the basic mercuric sulfate, or mercuric dioxysulfate, and the electrolyte is a zinc sulfate solution in which the cathode is stable. The anode is amalgamated zinc. The electrochemical system in the presence of an aqueous solution of zinc sulfate can be expressed as $Zn/ZnSO_4/HgO{\cdot}2HgSO_4$.

The system $Zn/ZnSO_4/HgO{\cdot}2HgSO_4 + C$ has an overall reaction on discharge of $3Zn + HgSO_4{\cdot}2HgO \rightarrow ZnSO_4 + 2ZnO + 3Hg$ and the potential of the cell is 1.35 volts. The theoretical capacity of $HgSO_4{\cdot}2HgO$ is 0.2204 ampere-hour per gram, and the practical capacity of the carbon mix is 0.15 ampere-hour per gram.

*Cell Structures.* While this system can be used in a standard Leclanché dry-cell structure, one of the important characteristics of the zinc/mercuric dioxysulfate electrochemical system is that it allows the use of stainless steel containers for contact to the cathode. This has permitted production of thin, wafer-type cells of the structure shown in fig. 26. The case is a shallow stainless steel

*Fig. 26. Wafer-type Sealed Zinc/Mercuric Dioxysulfate Cell*

cup; the cathode, a pressed pellet of mercuric dioxysulfate, Shawinigan carbon, and zinc sulfate solution containing 1 percent potassium dichromate as an inhibitor, in contact with the inner bottom of the stainless steel cup. The spacer is a laminate of paper and cellophane which separates the cathode from the amalgamated zinc disc anode. The anode has a polyethylene grommet around its edge, insulating it from the cathode container which is crimped against it for sealing of the cell. The electrical and dimensional specifications of the WD-5 wafer-type cell are 2.54 centimeters diameter, 0.272 centimeter height, 1.376 cubic centimeters volume, and 4.9 grams weight. Initial flash currents are on the order of 0.5 ampere; capacity is 230 milliampere-hour, with a 1000-hertz impedance of 10 ohms.

The present application of this cell is for special electronic wrist watches that function for two years before requiring replacement of the battery. The cell has not had extensive commercial application, possibly because of the dissimilarity between its dimensions and those of the miniature button-type cells. Its shelf life is excellent, cells of five years shelf time having maintained voltage and capacity without any electrolyte creepage or surface corrosion.

The motivation for the development of miniaturized batteries providing a higher potential per unit volume and capable of withstanding wide ambient temperatures derived from the need for operating military equipment. A source of potential was required for maintaining a continuous charge on a higher-grade electrostatic capacitor. The current flow needed for this application was in the submicroampere range. While the production of compact high-voltage piles was initiated by Zamboni in 1842, the Zamboni type of primary cell would not suffice for the operation of modern equipment and for the ambient requirements.

Batteries were developed in the Ruben Laboratories and were manufactured by Mallory Battery Company, a divison of P.R. Mallory & Co Inc., under the name "Solidion battery." The basic concept in the development of this battery was the use of potential developed between solid ionic conductors. Cells for the Solidion battery were made by compressing thin discs of lead peroxide and stannous sulfate to high densities. The potential developed at the junction of the discs was 1.56 volts. Contact with the couple was made with nonreactive metal discs such as chrome-plated copper, or nickel-alloy discs. After assembly in a stack with contact terminals at each end, the unit was incapsulated in an epoxy resin. This provided a unit capable of a high ratio of voltage to volume and weight with a long shelf life of five to ten years. These batteries also had a high resistance to impact and acceleration, with freedom from corrosive action.

*Fig. 27. Single 50-volt Stack of Solidion Cells*

Fig. 27 illustrates a single stack of 50 volts, 0.08 cubic centimeter in size and 7.3 grams in weight. Besides the use in military devices requiring the maintenance of a charged electrostatic capacitor, these batteries have been used in such specialized devices as Geiger counters and high-voltage calibration equipment and applied to circuits in which an electric charge on a capacitor must be maintained over a wide temperature range.

Unlike the so-called "dry cell" batteries of the Leclanché type, the Solidion battery is truly dry and hermetically sealed. The operation of the cell is believed to be the reaction of the $SO_4$ component of the $SnSO_4$ with the conversion of stannous to stannic sulfate.

The extremely small size of the individual Solidion Hi-Volt cells and the simplicity of their assembly lend themselves to variations in the geometry of the finished battery to meet specialized requirements. As a rule, these configurations can be estimated at 80 volts per cubic centimeter.

*Solid-State Couples*

Other types of solid-electrolyte batteries have been produced, such as cells composed of a silver-anode and a halogen-compound cathode. Some of these cells have a higher capacity than the Solidion cell and have supplanted it for a number of applications. In one of these batteries a series stack of solid ionic-conductive discs in contact with silver discs such as $Ag/AgBr/CuBr_3$ has been produced to provide a potential of 95 volts.

The use of alkali iodides such as $RbI_3$ allows a cathodic reaction when in contact with a silver anode, producing silver iodide. The overall reaction could be expressed as:

$$2Ag + RbI_3 \rightarrow 2AgI + RbI$$

The cathode terminal contact can be an inert material such as platinum or nickel alloy.

A number of solid state galvanic couples using alkali metals such as lithium have been developed with cathode reactants of conductive sulfides, selenides, or tellurides. The application has been essentially for use at low current density. Some implantable cardiac pacemakers utilize lithium anodes in contact with iron sulfides.

When high current densities are required, such couples as a lithium aluminum alloy anode and an iron sulfide cathode with a molten halogen salt electrolyte at a temperature of 400 to 450°C. may be used.

# CHAPTER VII

*Aerospace Requirements and Product Development—Cells having maximum energy density—Nonpolarizing organic electrolytes—Anodes providing high potential—Lithium-anode cells with solid carbon fluoride cathodic reactants—Sulfur dioxide cathodic-reactant cells.*

## *New Requirements for High-Energy-Density Cells*

With the advent of aerospace science, satellite communication, and marine exploration, the need for primary batteries with higher energy and voltage density has been recognized. To meet these requirements, cells have been constructed with lithium anodes.

Lithium has the highest electrode oxidation potential (3.045 volts), but it basically requires an organic electrolyte, such as propylene carbonate or acetonitrile, having a dissolved content of an ionizing salt such as one molar content of lithium perchlorate. The use of a lithium anode in an organic electrolyte also allows solution of the oldest battery or cell problem—that is, it achieves an electrochemical cell reaction without hydrogen polarization on the cathode. The cathode reactant or depolarizers must be insoluble in this organic electrolyte.

In order to make use of the higher anode potential of the lithium metal, various organic electrolytes have been formulated by dissolving soluble salts, such as lithium bromide or lithium perchlorate, in solvents, such as propylene carbonate, butrolactone,

dimethyl carbonate, ethyl acetate, nitromethane, dimethyl formamide, or tetrahydrofuran. The semiconductive cathode materials can be cupric sulfides or the fluorides of copper, nickel, carbon, and other materials. The use of gases, such as sulfur dioxide, as the effective cathode agent has found an application with a propylene carbonate/lithium perchlorate electrolyte.

The intercalation compounds of graphite and fluorine that have some practical production aspects are those of the general formula $(C_xF)_n$ where $x$ is a numerical value between 3.5 and 7.5, and $n$ reflects the presence of an infinite number of recurring $(C_xF)_n$ groups in the intercalation compound.

Lithium has been chosen as an anode for obtaining a high-energy battery since it has a desirable coulombic capacity. Its advantageous properties as an anode in comparison to magnesium and zinc in an oxidation reaction can be seen from the following tabulation:

| *Anode metal* | *Ampere-hour per kilogram* | *Watt-hour per kilogram* | *Kilogram per watt-hour* |
|---|---|---|---|
| Li | 794.69 | 2720.25 | 0.360 |
| Mg | 450 | 1290.50 | 0.71 |
| Zn | 167 | 276.30 | 3.59 |

The cathode composition is important in terms of the total coulombic capacity in ampere-hours per cubic centimeter.

| *System* | | *Reaction product* | *Ampere-hour per cubic centimeter* |
|---|---|---|---|
| $Li/(C_xF)_n = CF_n + nLi$ | → | $Cn + LiF$ | 1.769 |
| $Li/SO_2 = 2Li + 2SO_2$ | → | $Li_2S_2O_4$ | 0.77 |
| $Li/CuO = 2Li + CuO$ | → | $Li_2O + Cu$ | 2.40 |

In some experimental cells the electrolyte functions as both cathodic and anodic reactant. An example of this is the use of thionyl chloride ($SO_2Cl$) with end product of $Li_2S_2O_4$ + LiCl. With an organic electrolyte embodying propylene carbonate as the solvent, the electrolyte may have one molar strength of lithium perchlorate dissolved in it. An important result obtained with lithium anodes in organic electrolytes and a sealed carbon fluoride

cathode is the high voltage of 3.3 volts. The avoidance of hydrogen polarization allows discharges of high current densities, the extend of which is determined by the particular cathode composition used.

The cathode for use in the organic compound can be a compatible solid cathodic-reactant material, such as conductive oxides, fluorides, sulfides or other reducible compounds, with the addition of a finely divided conductor, such as graphite or silver powder.

The solid carbon fluoride compound $(C_xF)_n$ is prepared by reacting pure micronized graphite with a mixture of hydrogen fluoride and fluorine, the resultant product being washed free of acids and dried. To increase the cathode electrical conductivity, the fluorides are mixed with about 10 percent by weight of micronized graphite. The mix is pressed at about 344.89 kilograms per square centimeter on to a nickel metal screen. The anode is made by rolling lithium rods to a thin strip in an inert atmosphere, then pressing them into and on a copper screen.

Cells of $(CF)_n$ cathode-reactant type have been reported to have coulombic efficiencies ranging from 70 percent to 92 percent, dependent upon the current density of the discharge. The $(CF)_n$ is efficiently reduced to C with a capacity of 480 watt-hours per kilogram:

$$nL + (CF)_n \rightarrow LiF + nC$$

Cells of the lithium/carbon fluoride type have an open circuit potential of 3.3 volts.

The solid cathodic-reactant electrodes do not have as wide a range of operating temperature as cells using $SO_2$ as the cathode reactant. The gas under pressure provides a continuously replaceable active reactant at the cathode surface without accumulation of impeding end products.

One of the $SO_2$ type lithium cells being manufactured currently by Mallory uses a cathode/anode structure similar to that used with the $(CF)_n$ cell with an electrolyte of propylene carbonate or acetonitrile and an ionizing salt content. Its lithium anode is composed of a lithium sheet pressed into a metal screen. The cathode is a carbonaceous mixture, such as, for example, graphite or acetylene

black, and a binder pressed into a metal screen. The assembly is placed in a nickel-plated steel container. All of the important features of the cell are illustrated in Fig. 28.

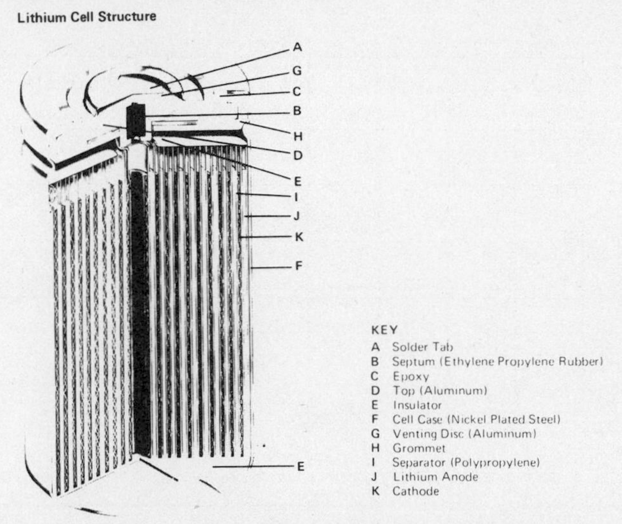


*Fig. 28. Wound-electrode Lithium-anode, Organic Electrolyte Cell*

The cathodic reactant, or depolarizer agent, is a high-pressure atmosphere of sulfur dioxide (3.5 atmospheres at 25°C.) in contact with the carbonaceous screen. Since there is no hydrogen to contend with during or after discharge, the cell has a vent for the purpose of taking care of extensive internal gas pressure that might be caused by overload or short-circuit conditions.

The nominal voltage of a lithium/$SO_2$ cell is advertised as 2.95 volts, with operation over a wide range of temperatures from −40°C. to 70°C.

Where a gas, such as sulfur dioxide, which is 71% by weight of the electrolyte, is utilized as the cathode reactant, it can be added to the cell at low temperature in a liquified state. The cathodic reaction takes place between the interface of a graphite-filled screen and the electrolyte. The overall reaction can be stated as follows:

$$2Li + 2SO_2 \rightarrow Li_2S_2O_4$$

The electrodes are spirally wound with a separating spacer, as shown in Fig. 28, and the unit is sealed in a metal container.

A number of practical production problems remains to be resolved before the lithium-anode cell can become a large-scale competitor to the zinc-alkaline primary cells.

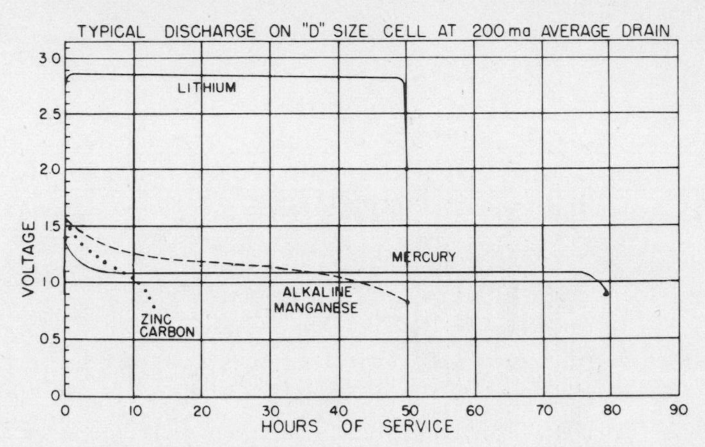


*Fig. 29. Comparative Discharge Characteristics of Lithium and Zinc Anode Cells*

*Review of the Evolution of Primary Cell Technology*

The electric cell has come a long way since the recognition of Volta (1800) of the electromotive force generated when dissimilar metals are in contact with, and separated by, an aqueous chemical solution. The evolution of the primary cell could be reviewed, perhaps simplistically, as follows:

Progress was made by Davy in meeting the inherent problem of cathodic hydrogen polarization—which limited the quantity of current flow generated by the oxidation potential of the zinc anode—when he utilized nitric acid as the electrolyte. This electrolyte, together with a carbon cathode, produced a cell with a longer discharge time at higher current by virtue of its hydrogen-oxidation character. The single-fluid electrolyte, however, entailed an equally limiting practical problem: namely, the accelerated corrosion of the zinc anode and self-discharge by localized reactions. The apparent partial solution to this secondary problem was to withdraw the active electrodes mechanically from the corrosive electrolyte during nonuse in the manner devised by Wollaston.

To meet the electrode-corrosion problem more realistically, the use of dual electrolytes was developed; in the work of Grove and Bunsen, a more oxidizing electrolyte, such as nitric acid, contacted the cathode in a porous cup compartment that was immersed in a low concentration of sulfuric acid containing the zinc anode. Both the anolyte and catalyte portions of the electrolyte were chemically compatible, with excellent ionic exchange between them through the separating porous cup. Improvement of this cell system came about when Bunsen substituted carbon for Grove's platinum as the cathode electrode.

An important improvement was made by Poggendorff when he substituted chromic acid, derived from a mixture of sulfuric acid and sodium bichromate, for nitric acid as the catalyte, eliminating the production of hazardous nitric oxide fumes.

The application of Kemp's proposition that there would be less corrosion of the zinc anode if it were amalgamated with mercury was fundamentally helpful for all zinc-anode primary cell systems.

Daniell developed a two-fluid noncorrosive electrolyte system with a separate anolyte and catalyte, thereby avoiding the problem of hydrogen polarization on discharge. He utilized noncorrosive electrolytes separated by a porous cup that discharged copper ions instead of hydrogen ions to the cathode. The copper ions deposited nonpolarizing copper on the cathode. The zinc sulfate in the anolyte compartment containing the zinc anode provided the ionization necessary for generation of the cell potential. Varley's modification of the Daniell cell to a gravity-separated electrolyte structure materially reduced the internal resistance of the cell and made it more viable for many applications, particularly for the telegraph industry.

In response to the requirements of railroad signal systems for higher-energy-density cells capable of operating under wide variations of ambient temperatures, Lalande and Chaperon employed the reduction reaction at a cupric oxide cathode in a single-fluid alkaline electrolyte and an amalgamated zinc anode. Edison's improvement in the cupric oxide electrode composition helped to make the Lalande system more practical.

The invention by Leclanché of a single-fluid electrolyte system of ammonium chloride and the use of a solid insoluble manganese dioxide cathode reactant produced a higher-voltage cell (1.5 volts) than that of Daniell (1.08 volts). Gassner's modification of the Leclanché system to an immobilized electrolyte and a more portable dry-battery structure, with the anode also serving as the container, helped materially to expand the Leclanché system cell to worldwide use.

The development of the sealed zinc/mercuric oxide alkaline cell by Ruben during World War II provided a method for the miniaturization of cells where high coulombic capacity in smallest size and constant load voltages within prescribed limits were desirable. The substitution of silver oxide ($Ag_2O$) for mercuric oxide has increased the potential of the sealed alkaline cell. Development of the zinc/silver oxide and cadmium/silver oxide alkaline rechargeable cells supplied a high energy density battery for early aerospace and marine applications. The cadmium/nickel oxide alkaline cell, with its better cycleability and lower cost, has supplanted the silver oxide cells in many applications. The relative watt hours per kilogram of these systems are as follows:

| | *watt hours per kilogram* |
|---|---|
| zinc/silver oxide | 133.2 |
| cadmium/silver oxide | 60 |
| cadmium/nickel oxide | 26 |
| iron/nickel oxide | 24.2 |

# CHAPTER VIII

*Secondary batteries—The battery requirements of the automobile industry—The rechargeable battery—History and development of the lead/acid cell—Planté cell in 1860—Faure cell in 1881—Brush and Volckman's structural change, providing pockets in lead plate for support of lead oxide—Improvements in shelf life and capacity per unit weight by material and structural changes—Experimental cell with expanded titanium-positive-grid electrode and expanded stainless steel negative grid with titanyl sulfate-buffered sulfuric acid electrolyte gelled with fumed silica—Practical requirements for vehicle motive energy—Special rechargeable alkaline batteries, such as zinc/silver oxide—Replaceable-anode primary zinc/air-depolarizer cells—Present and future storage battery requirements.*

*Lead/Acid Rechargeable Batteries*

The principle of the lead/acid rechargeable battery was discovered by Raymond Gaston Planté while studying the polarization effect on metal electrodes immersed in sulfuric acid when current was discharged between the electrodes. At a meeting of L'Académie des Sciences, March 26, 1860, (Compt. Rend 50, 640 [1860]) he reported that after disconnection from the external charging source, the lead plates would discharge a current, the extent of which was dependent upon the time and current from the source of charging current. The charging of the spaced lead-plate couple in the sulfuric acid produced an integrally formed layer of lead peroxide on the positive electrode. Planté applied the term "accumulator" in 1879, in view of his experience in producing a chargeable lead/acid cell in which the charge was chemically accumulated.

He reported that the longer he applied the oxidizing current to the charging or positive plate, the thicker would be the integrally formed lead peroxide layer on the positive plate and the greater would be the discharge capacity of the cell when connected to an external load circuit.

He noted the limitation in depth to which he could form an oxide layer; to increase the discharge capacity of his rechargeable cell, or battery, to provide a higher capacity, he increased the area of the lead-plate electrode. He made large-electrode-area cells by means of a helical winding in the manner of Hare's primary cells. His cell comprised two lead-strip electrodes separated by linen-cloth spacers, helically wound to give the maximum area, immersed in a jar of sulfuric acid. One of the limitations experienced by Planté in developing rechargeable batteries with increased size and capacity was the fact that he had to use primary cells, such as the Bunsen cell, for recharging. With the development of the direct-current generator, batteries could be constructed having high storage capacity, thus overcoming one of the limitations imposed on his early work.

However, Camille Faure, in 1881, recognized that the limitation in obtaining maximum cathodic capacity in a battery for a given volume was determined by the limited lead peroxide formation by electrolytic oxidation at the positive electrode.

To overcome this limitation, he coated the lead plates with a paste made by mixing lead oxide with sulfuric acid, which he allowed to dry to a hard coating. He then used this as the positive plate in combination with a spaced lead electrode as the negative plate in an electrolyte of sulfuric acid. On passage of current between the plates a cathodically active thick layer of lead peroxide was formed. Any lead oxide or sulfate on the anode was reduced to a porous lead coating.

The Faure oxide-coated lead plate provided a large increase in capacity over the electrolytically oxidized lead plates of Planté although poor oxide adherence with use was a practical limitation. Brush, in America, prepared active plates by mechanically bonding the lead oxide to the lead plates and obtained patents which were a dominating factor in this field. In Europe, Volckman perforated the lead plates to provide pockets for support of the oxide.

For maximum electrochemical cell efficiency, it is desirable to have minimum electrical resistance between the lead electrodes and the cathodic or anodic reactants and the electrolyte. Since the resistivity of lead peroxide is ten thousand times that of lead, the electrical path length and contact resistance of the oxide to its supporting lead base should be as low as possible. The application of cast-grilled lead plates with lead peroxide in their retainer spacings provided the practical means for more widespread application of the Planté lead/acid battery (fig. 30).

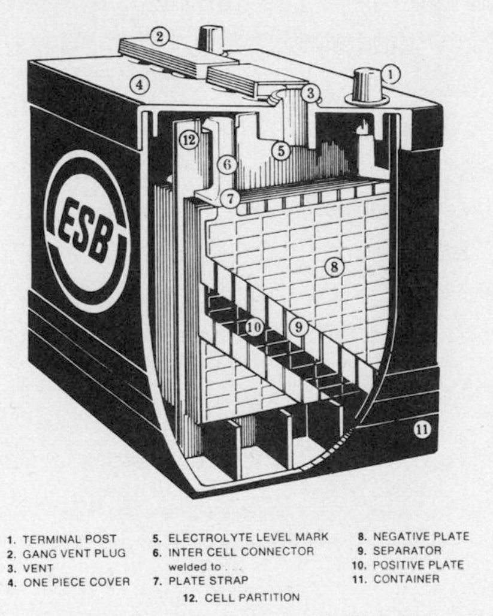


*Fig. 30. Structure of Commercial Cast-grid Lead/Acid Storage Battery*

The development of improved electric generators for recharging rapidly launched the lead/acid storage battery into many industrial uses. In the United States, a central station battery was installed in Phillipsburg, New Jersey, in 1885, only four years after the development of the pasted-lead/acid cell. In Dover, New Hampshire, batteries were used in the operation of a trolley car. In 1888, on the trains of the Pennsylvania Railroad, batteries were installed on cars for train lighting service; and in isolated power plants in Baltimore, Maryland, batteries were used to maintain capacity or load-leveling of a central service station. In 1889 batteries were employed in Chicago, Illinois, for central telephone service.

With the development of an improved and more portable structure in 1900, the lead/acid battery was applied to many other mobile applications, such as submarine propulsion, electric vehicles, lighting for automobiles, heavy-duty railway signal circuits, and the self-starter for the internal combustion engine.

In the 1920s the lead/acid battery was the source of electric energy for heating the filament cathode of vacuum-tube radios. Emergency radio and lighting equipment on steamships depended upon the relatively high-energy capacity of this storage battery as a reserve source of power.

At the turn of the century, three types of self-propelled vehicles were being made, the steam car, the gasoline car, and the electric car, each striving to be so superior that it would capture a major part of the market. Despite the lower range of the electric car, it was the one preferred by women, since the hand-cranking of the internal combustion engine for starting was an arduous and sometimes hazardous task.

Means to start the internal combustion automobile engine by compressed air motors were devised. However, the introduction of the electric motor starter, operating from a lead/acid storage battery, eliminated the problems of starting the gasoline engine and created a worldwide demand for lead/acid storage batteries. Following the application of batteries, a method for recharging them was found in the direct-current generator with voltage control relays. These recharging D.C. systems were supplanted many years later by the A.C. generator and the solid-state rectifier that provided a maintenance-free charging system.

*Internal Operation of the Lead/Acid Cell*

The electrolyte of sulfuric acid is the ion transfer agent for the charge and discharge of the cell. The sulfuric acid electrolyte contains two hydrogen ions 2 ($H^+$), each carrying a charge of an electron, and a sulfate ion $SO_4^{--}$ lacking two electron charges, thus giving two positive $H^+$ ions and one negative $SO_4^{--}$ ion. When on discharge, the sum of the oxidation and reduction potentials of the lead and lead peroxide electrode (2.04 volts) causes the flow of the hydrogen ions to the positive plate ($PbO_2$), giving up their charge, and in the presence of the sulfuric acid, reducing the lead peroxide to lead sulfate and releasing free water:

$$PbO_2 + 2H^+ + H_2SO_4 \rightarrow PbSO_4 + 2H_2O$$

Similarly the iead negative plate (Pb) receives the positive ion $SO_4^{++}$ with production of lead sulfate:

$$Pb \rightarrow Pb^{++} + SO_4^{--} \rightarrow PbSO_4$$

The overall result of the two ionic reactions of the reactants may be stated:

$$PbO_2 + 2H_2SO_4 + Pb \rightarrow 2PbSO_4 + 2H_2O$$

The charging is the reverse action of the ionic components of the electrolyte on the insoluble lead sulfate. One equivalent of $PbO_2$ and one equivalent of Pb are modified by the passage of one faraday. Since the potential of the lead acid cell in operation is a function of its electrolyte specific gravity, the cell voltage may be expressed as:

$$E = 1.85 + 0.917\,(G\text{-}1)\text{ volts}$$

where G is the electrolyte specific gravity.

Lead acid cells have a temperature/voltage coefficient in the order of 0.000395 volt per degree Celsius.

The weight of the reactants $PbO_2$ and Pb and $H_2SO_4$ in a lead/acid cell constitutes an inherent limitation in the possible capacity per unit weight of the system. This in part illustrates the limiting factors for obtaining low-weight batteries for vehicle-traction use where kilowatt-hour per pound or kilogram is the basic factor. To this must be added the fact that in some lead/acid batteries the inactive lead-metal support grid and connectors may constitute 50 percent of the plate weight.

The problems in supplying electric current for the electric car involve the basic requirement of adequate energy to operate the motor vehicle system over a practical distance. The battery should have adequate capacity to allow the vehicle to travel a distance of several hundred kilometers with a realistic battery weight. The comparison has been made to a gasoline engine and its fuel supply: a 75-liter tankful of gasoline, weighing 54 kilograms and occupying 0.07 cubic meter, gives a car at 15 miles per U.S. gallon (6.3 kilometers per liter) a range of 300 miles (480 kilometers). The efficiency of the gasoline engine is about 30 percent under good driving conditions.

Batteries for electric-car use would have to be evaluated comparatively in terms of watt-hours per kilogram, recharging, and replacement cost. The average lead/acid battery has a capacity on the order of 30 to 40 watt-hours per kilogram, with some increase in capacity per unit weight with use of lower density and thinner grid materials. In a typical lead/acid cell the reactant-supporting lead grid weighs from 35 to 50 percent of the pasted grid weight.

One ampere-hour equates to 3.866 grams of Pb and 4.463 grams of $PbO_2$, giving a total of 8.329 grams of active material per ampere-hour. The use coefficient is in the order of 50 percent, so that it requires 16.65 grams of active materials per ampere-hour.

One limitation for maximum capacity is in the positive electrode. An increase in efficiency and capacity should come with increased study of lead peroxide characteristics and further knowledge of its physical chemistry. Increased quantity of active lead peroxide on the positive plate is not always the answer, since such factors as ionic polarization and ohmic resistance with long-path lengths are important.

The development by the Electric Storage Battery Co. of tubular-plate batteries, in which the lead peroxide is packed into porous glass-cloth tubes with internal lead- or alloy-rod contacts, or conductive spines, has provided efficient means for constructing high-capacity traction-type batteries with a lower inactive lead content.

Some of the limitations involved in design of electric cars using lead/acid batteries of the type now used for commercial trucks,

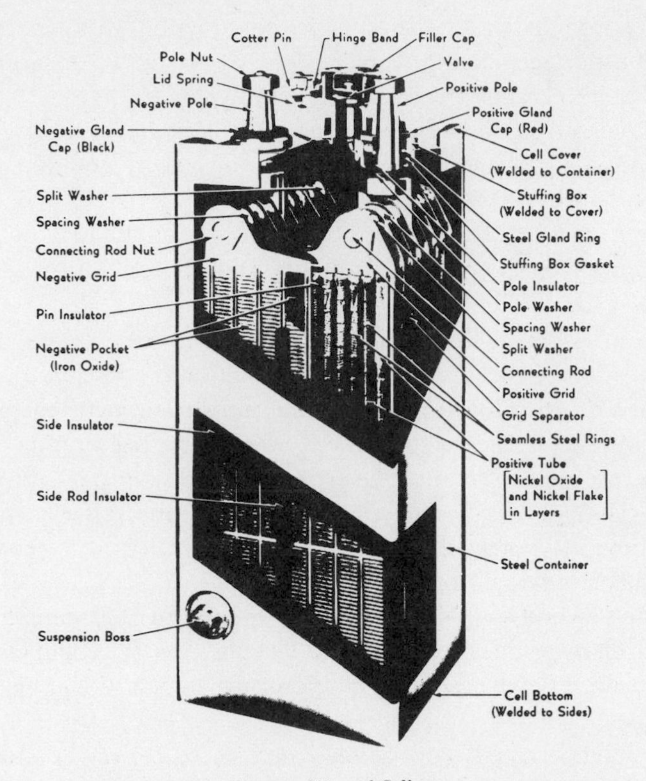


Cutaway View of Cell

*Fig. 31. Porous Tubular-Cathode Nickel/Iron Alkaline Battery*

buses, farm machinery, fork-lift trucks, railroad trains, and other heavy equipment can be envisioned from the following reported calculations.

A small car like the Volkswagen would require a 15-horsepower electric motor for a 240-kilometer range; employing present-type lead/acid batteries, it would weigh 315 kilograms more than the completely equipped present model VW. The initial battery cost, together with the replacement cost, since the battery is limited in cycle life, would make it economically undesirable.

At the present time considerable effort is being devoted to hybrid systems, such as the combination of lead/acid or nickel/iron batteries operating in conjunction with an internal combustion engine. Research could also supply a means for electrically charging the battery at various stages.

Imaginative thinking with modern technologies will be required to translate the conceptual possibilities of battery-powered electric automobiles into realities.

Perhaps a hybrid system utilizing the combination of stored mechanical energy coupled with electric drive could have possibilities as a low-weight hybrid power system.

The flywheel is an efficient energy-storing device, and by the application of the modern science of materials and engineering, it may be made to function as a partial supply of kinetic energy to a hybrid system. The capacity of a flywheel increases with the square of its top speed, so that by the proper choice of strong, light-weight materials and friction-reduction means for coupling this energy to a common shaft of an electric drive, it is possible that a lower-weight hybrid drive system could be produced.

The flywheel could be charged by rotation to high velocity from a stationary engine by a cable, and its kinetic energy could be fed to the drive of a battery-driven motor when additional power is needed.

A good perspective on the recognized need for the development of a practical storage battery system capable of meeting traction requirements and some of the early history of those engaged in producing an electric automobile is contained in the following memorandum sent to the writer in March 1967 by Mr. Frank Babcock of New York City, a pioneer in electric cars.

*Memo on "Electric-powered Cars"*
(received from Frank Babcock, March 1967)

The recent publicity given this subject has been of great interest to me in view of the fact that, together with my father, I was identified with the early development and marketing of electric vehicles.

Operating first as The Buffalo Electric Carriage Company (later with corporate name changed to Babcock Electric Carriage Company), in March 1899, we began production on a really "Horseless Carriage" model designated as The Babcock Stanhope. As my father had previously manufactured an extensive line of horse-drawn carriages in Amesbury, Mass., under the name of The F. A. Babcock Co., it was quite natural that the first model in this new field should more or less follow construction and design of the carriages which had been so well received.

Thus, the Stanhope, minus its horse-drawn "shafts," was decided upon for the first model. Its "power plant" was based on a design furnished by T. A. Willard of Cleveland, Ohio, who, at that time was manufacturing storage batteries under the name of Sipe and Siegler (later to become the Willard Storage Battery Co.). His interest, of course, was to promote the development of the Electric Vehicle as an additional outlet and market for his company's products. At the same time, he had promoted the development of a model for Riker Motor Vehicle Co. and for Walter Baker in the design of the Baker Electric "Piano-box Buggy" model.

At that period, the only type of storage battery available for such application was what was known as the Planté. Of very heavy lead plates, limited amp. hr. capacity per lb., this power-pack of 40 cells (80 volts) would only provide about 35-40 miles per charge under good road conditions in our first production.

Even with its limitation of speed and mileage at that period, the Electric was quickly accepted by those who couldn't well cope with cranking and the eccentricities of the gas engines then available. Though we exhibited at the first automobile shows at the original Madison Square Garden and at Grand Central Palace in November 1899 and sold about 100 units, we recognized that, if we were to progress, it would be necessary to attain better speed, mileage per charge and redesign in over-all matters.

It was shortly after these shows that an entirely different battery was presented to us by Henry Porter of Waukegan, Ill. The Porter Battery Co. was producing cells with very thin, "pasted"-type plates, known as the Faure, with amp. hrs. capacity per lb. far ahead of the Planté and which supplied far more mileage of course, also at a price much below the Planté.

Of course, we adopted this battery and were therefore able to surpass the performance of our competitors until, in November 1902 we were restrained, thru a court order obtained by the Electric Storage Battery Co. of Philadelphia, Pa., citing infringement by us under the Brush Patent owned by them. The Brush Patent was to expire March 1903. We had, meanwhile, found that Willard and the Philadelphia Storage Battery Co. (later Philco) were in a position to supply us with excess requirements which were increasing due to our growth.

Perhaps right here is an appropriate spot to explain our battery set-up after our "initiation" into all these angles. First, we had sought and obtained through Porter, a lighter and higher capacity type battery that, even with shorter life, would give greater dollar value and performance than any other available at that time. Second, as our requirements increased and it became evident that there was a profit to be made in these batteries if we bought the "raw materials," so to speak, we made deals with Willard, Philco and Porter to furnish us the positive and negative elements with primary formation of the plates. These were put into hard rubber jars with the Babcock name on them, and discarding the rubber separation generally used, were replaced by apple-wood separation to reduce internal resistance. Then, completing the formation cycle of the elements, we assembled the cells in wooden trays built in our body shop. The complete battery was then ready for installation in a new car or for replacement of one that was under-capacity. Our Guarantee was that we would replace any of our batteries every 12 months with a new one for $100.00. This was a cost per month of $8.66 with the assurance that, with its high capacity, it would give more speed and mileage than any other battery and at very much less cost per month than the heavy, thick plate Exide or any other then in use by our competitors.

Well—as this "extra mileage—higher speed" requirement increased in our market demands, we met the situation by a redesign of our models. From high carriage type wheels with solid rubber tires, short wheel-base of carriage shop construction and a tiller handle for steering, we began to adopt some features from the gas-powered cars, then offering a bit more than the "Electrics"—except that the ladies were still our prospects as they couldn't well crank the engines.

Our "Victoria" with its low-hung body, handsome leather top and fenders to protect the long skirts from the pneumatic tires made an immediate hit. However, strange as it may seem now, our innovation of a steering wheel to replace the "stick steer" was met, at first, with many women, that it would look "too sporty" for ladies. We had little difficulty in overcoming this when we stressed that it was there as a safety feature for better control and that, of course, the Babcock with a top speed of 35 m.p.h. was confronted with the same road hazards as the "gas-buggy." As it cost us $28.00 more to put a wheel steer on a car, we would offer to replace it in two weeks if the lady would prefer the lever. As I remember it, we were never called on to replace one. Later on we even went so far as to provide for *tilting the wheel* to allow more ease in seating. A feature now being exploited in many cars as something quite new and novel.

One model, of which only one was built, as was a "rear entrance tonneau" touring type car. Four passengers, no top, right-hand steering wheel, with control lever under the wheel as is done today. Equipped with motor drive on each rear wheel thru spur gears and a 40 cell, 350 amp. hr. battery distributed front and rear, wheels of the artillery type, solid rubber tires with a steel rim on each wheel that allowed travel on a standard gauge R.R. or trolley track. I drove this vehicle from Boston to New York in November 1903 as a publicity stunt. Took three days on the gravel roads then in use and provided charging at any sources available—which in those days were scarce. One being from a 220 H.P. Corliss steam engine driving a 220 volt generator for which I paid $7.00 for the coal used in a 3 hour charge. Used the inter-urbans rail once or twice to get off the bad road stretches, much to the annoyance of the motor-man.

Some of the other models developed were, of course, the Broughams, Coupe, Roadsters and even a taxi-cab. Also another design of a touring car. We enjoyed a successful and prosperous business until 1913 when the "self-starter" on gas cars made them more practical and acceptable for women to handle and to enjoy the advantage of greater mileage and also at a lower first cost than that of the Electrics.

It was in 1912 that we sold our assets to a new company, The Buffalo Electric Vehicle Co., which was formed by a merger of the Babcock Co., a body company and the Clark Electric Co. However, thru a combination of declining market for Electrics and extremely poor management, they folded up in a year or two.

The only companies that continued on after that were The Rausch and Lang Carriage Company, Baker Electric Vehicle Co. and The Detroit (Anderson) Electric. R. & L. were taken over by Baker, and then Baker concentrated on building industrial trucks as it still does today as a unit of The Otis Elevator Co.

I represented the Baker Co. in Buffalo for a few years after we sold the Babcock Co. and subsequently joined their New York representative in promoting their electric industrial trucks and continued with them until 1960 under The Otis Co.

With this background, it is quite natural that the recent publicity given to the probable "rebirth" of the Electric Vehicle should stir up my interest in the subject. Especially so as I expressed my firm convictions in this respect during a visit with Mr. Thos. Edison at West Orange, N.J., in 1912. In a discussion of the limitation of his and all other storage batteries, I told him that he probably wouldn't live to see it (nor did I expect to) but that "someday" the storage battery as we then knew it would be "as dead as the dodo bird" and quite obsolete. That, with the developments going on in the Chemical, Metallurgical and Electrical Engineering fields, there would be a "Primary Source of Power" developed for transportation and many other fields and use quite apart and different from the storage battery.

Well, here I am still around and this conviction is still with me. It's not far off today. I can't quite go along with the "fuel cell" idea nor, to date, have I learned of what I consider the right answer in the various other batteries being offered. Jokingly, I've often said that when they can harness some form of atomic energy or develop a cell to provide energy by dropping a "pill" into an electrolyte to get away from dependence on a charging source—"that will be the day."

The present-day so-called "dry cell," such as used in transistor radios, even though lacking the characteristics required for heavier duty use—excessive current demand, substantially higher per-cell voltage and greater amp. hr. capacity—is basically what I should like to see given more attention and research. A true Primary Source of Power with perhaps 1000 hrs. of practical life, then replaced at a cost that is reasonable.

Anyway, I have every reason to believe that the rebirth of the Electric Vehicle is not far off and I hope to be around to witness its contribution to cleaner and purer air, less noise and taking its place in this Electronic World.

Other primary-cell secondary battery systems are presently being explored for traction purposes, such as zinc air cells and zinc chlorine cells. In these cells the cathode reactants are gas, with cathodic reactions occurring at the interface of the cathode terminal electrode, which may be carbon or graphite, and the cation flow to the interface of electrolyte and cathode providing the site for deionization.

The spent-anode zinc may be recovered to some degree, in situ, by recharging of the cell, or it can be externally reclaimed by electrodeposition from spent electrolyte in the cell. With alkaline electrolytes the recovery of anodic zinc from the zincate electrolyte is difficult to accomplish since the tendency is for zinc ionic concentrations that cause dendritic nonuniform zinc deposition. This is less a factor with acidic electrolytes.

The use of chlorine or bromine as cathodic reactants offers a promising possibility for high-energy-density cells that may serve for automobile or load-leveling applications. The use of chlorine as a cathodic reactant was described by Upwand in 1876. He used zinc chloride as the electrolyte and carbon as the cathodic electrode. The potential of the Zn/ZnCl aq./$Cl_2$/C cell is 2.1 volts, with the reaction end product of $ZnCl_2$ aq.

The basic problem for traction use remains one of cost and ratio of battery to vehicle weight. However, under the pressure caused by financial and political factors inherent in our petroleum supply, a motivating factor is introduced that may well accelerate the time of conception and development of a realistic system. Numerous exotic systems have been proposed, but it will take time to prove their economic feasibility and safety.

*Lower-Weight Lead/Acid Batteries*

One of the problems associated with the large weight factor in portable applications of lead/acid batteries is that of the weight ratio of inactive to active materials. One means of decreasing the percentage of inactive grid metal was a cell developed over several years and described in a patent (no. 3,798,070 [March 1974]) by Ruben. In this development, an expanded, thin titanium metal-grid electrode is used for support of the lead peroxide instead of a heavy lead metal grid. Titanium has a density of 4.3 grams per cubic centimeter as compared to 11.342 grams per cubic centimeter for lead at 20°C., and allows a lower weight support for the reactants. When so used, it is expanded and stretched to increase the holding depth of the uprighted strands.

Its high tensile strength allows the use of thinner metal with larger areas than is possible with lead. It also supplies a more uniform electrical contact to the reactants and is capable of meeting the density changes of the reactant in the charge and discharge process.

However, to utilize titanium metal as a grid requires two necessary processing steps. The first is to avoid polarization, on charging, between the surface of the titanium and the lead peroxide reactant when the grid is the positive terminal. This is accomplished by plating an integral flash coating of nickel followed by a dense lead plating from a lead fluoborate electrolyte. After drying and baking, the electroplating will avoid an interfacial polarized film-forming potential between the grid and the reactants ($PbSO_4$ and $PbO_2$) on charging. The grids are coated with a paste of lead oxide in the manner applied to lead plates. Another important requirement is that the sulfuric acid, when mixed with submicro-size formed silica to form the gelled electrolyte, must be buffered with a small percent of the basic titanium sulfate ($TiSO_4$) so that any titanium in contact with it becomes passive. Ruben found that the addition of a small amount of titanyl sulfate combined with the sulfuric acid prior to gelling, will protect the titanium on shelf against dissolving and forming titanium sulfate.

Titanium metal that would completely dissolve at 50°C. in sulfuric acid of 1.3 specific gravity in twenty-four hours will have a minor loss in weight in the titanyl-buffered solution over several years' time. Batteries on test for several years have maintained their conductivity and mechanical form within practical limits.

The negative grid requires a metal other than titanium, since titanium under pressure of hydrogen readily forms a hydride and becomes quite brittle. It was found that stainless steel grids, nickel and lead plated, also become passive in contact with the titanyl-buffered sulfuric acid gel. The use of a gelled electrolyte in a confined retainer reduces gravimetric separation of electrolyte when acid concentration changes occur on charge and discharge.

When the grids pasted with the lead oxide titanyl-buffered sulfuric acid have been processed and charged so that the positive grid paste is converted to $PbO_2$ and the negative to porous Pb, they are assembled into a unit. In order to have a recombination factor in a sealed cell, it is important to have an anode area larger than that of the cathode. For each lead peroxide-filled grid, there is a porous lead-filled grid facing each side.

The units for an 18-ampere-hour battery have a grid dimension of 84 square centimeters, made by assembling three negative expanded stainless steel base grids electroformed with porous lead and two lead peroxide-filled processed titanium grids, the plates being separated by two spacers, both of which are filled with the silica gelled electrolyte. The first spacer is a porous unwoven glass fiber of about 0.6 millimeter thickness, which when filled with gelled electrolyte is applied to the positive electrode. The second spacer is a polyethylene frame about 2.5 millimeters thick, the open space of which is filled with the gel. When the grids and spacers are assembled they are tightly wrapped with thin Saran film and sealed on all exposed ends by an acid-resistant adhesive tape.

The unit is then placed in a plastic container which is filled to the desired level with a warm acid-resistant epoxy resin, about 1 centimeter of the resin being above the cell unit. In a short time polymerization and hardening occur, with the result that the unit is encapsulated in a hard, solid block. The heat of reaction of the hardening resin helps to eliminate absorbed gases in the formed grids. After a sufficient aging time to ensure maximum strength of the resin, the unit is discharged to a low voltage at a 12-hour rate. It is then recharged and ready for application. The accompanying illustration (fig. 32) illustrates some of the cell parts and an encapsulated unit.

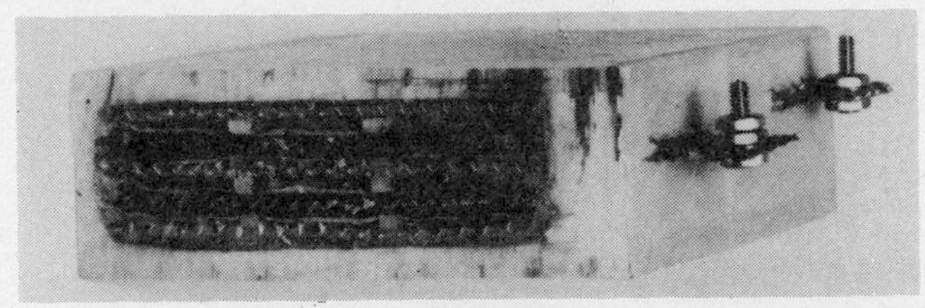

*Fig. 32. Expanded Titanium-positive Grid in Resin-encapsulated Gel Cell*

The combination of close spacing of the active electrode surfaces with larger anodic area and the mechanically solid encapsulation probably explains why, under prescribed charging, there is practically no gas-polarization voltage, so that a maintenance-free structure results, allowing continuous float or trickle charge operation. Large-scale tests over a long period will be required to evaluate the practical character of this development and its possible application to large-capacity storage batteries.

In view of the comparatively high price of titanium, development work is presently being carried on at the Ruben Laboratories to utilize expanded stainless steel for both positive and negative grids.

The standard lead/acid cell has been materially improved—to a structure requiring less maintenance—by the use of lead grids of higher purity and by the substitution of calcium for the antimony as a strengthening element for the lead. Future improvements in the lead/acid cell require a better understanding of the chemistry and physical chemistry of the lead peroxides and the possibilities of lead alloys for producing a better reversible lead peroxide cathodic reactant and the reduction of the irreversible form of lead sulfate.

The development of a traction battery that would have higher capacity per unit weight and volume than the lead/acid cell was actively undertaken by Junger in Sweden. He believed that the nickel oxide-cathode alkaline cell would provide the necessary properties for adequate energy density and rechargeability.

Desmazures, in the late 1880s, tried to convert the Chaperon-Lalande cell from a primary cell to a secondary battery and mentioned in one of his early patents the use of nickel oxide in place of the cupric oxide as a cathodic reactant. Junger in his early work had found that nickel hydroxide was a suitable active cathode material, and consistently held to his opinion of its advantageous cathodic capacity and stability.

He utilized the ferro-compounds as the anodic reactant, and he used graphite in both electrodes to improve the conductivity. He had experimented earlier with cadmium as an anodic material with cupric or silver oxide as cathode in an alkaline cell. With the improvement in production of a porous cadmium electrode, he adopted cadmium as the preferred metal in combination with nickel oxide. His utilization of a nickel metal support for both reactants helped materially to make a more practical cell.

Since Junger and Edison were independently engaged in the development of a nickel oxide rechargeable cell at the same time, there was a conflict in patent interests. Edison, who in the early 1890s improved the Chaperon-Lalande alkaline cell with his development of the bonded copper oxide cathode, tried, as Junger did, to convert it into a rechargeable cell useful as a traction battery. He did not believe it was practical to make a lead/acid battery suitable for small vehicles.

Edison's first nickel/iron alkaline storage battery utilized graphite for electronic contact to the nickel hydroxide. However, he found that the graphite which he was using was responsible for disruption of the reactant mass on cycling. By substitution of flaked nickel, he was able to make a more commercially suitable battery. The nickel oxide was mixed with nickel flakes and pressed into the reactant-container area of the plate. He also found that by adding mercuric oxide to the iron power of the anode and by using nickel flake in the cathode he had improved the efficiency, stability, and cyclic performance of his batteries. He found that he could advantageously use nickel-plated steel as his reactant-retainer metal, and the addition of lithium to the potassium hydroxide electrolyte afforded improved cyclic operation in applications where lead/acid batteries have been used.

Nickel/iron batteries have been in use for many industries, and today there is a resurgence in activity toward improving them for possible use as traction batteries for small electric-powered cars.

The normal overall reactions of the nickel/iron cells could be expressed as follows:

$$2NiOOH + H_2O + Fe \rightleftarrows 2\,Ni(OH)_2 + Fe(OH)_2$$

The anode reaction:

$$Fe + 2OH \rightarrow Fe(OH)_2 + 2e^-$$

The larger-size industrial alkaline batteries are manufactured in an unsealed structure containing a pocket-plate-type of electrode (fig. 33). Perforated nickel-plated steel is used to contain the active reactants with the electrode. The nickel hydroxide has flaked nickel to increase its conductivity and porosity. The smaller-sized units, with a capacity in the order of 33 watt-hours per kilogram, are constructed with sintered pressed-nickel-powder electrodes.

The sintered nickel-powdered electrode allows impregnation of the cathodic or anodic reactants in depth. A porosity of about 80 percent allows adequate reactant contact to the porous walls of the electrode structure. The surface area of the sintered electrode is in the order of 25 square meters per gram of nickel. This assists in increasing the cyclic charge/discharge characterisic, since there are minimum reactant-density changes from the reversible charge/discharge action. The higher overall conductivity of the active electrodes allows high discharge current density.

Considerable research is being carried out today to determine the best methods of using iron as the anode. Both Junger in Sweden and Edison in the United States explored the use initially at the turn of the century. Both considered the nickel/iron battery system as a possible answer to the need for a vehicle-traction battery. They both used cadmium as the anodes, but their initial results favored iron. Junger and his associates concentrated on the cadmium anode with its widespread application to uses other than providing traction.

*Nickel/Cadmium Battery*

Nickel hydroxide is the active material of the positive plates in the nickel/cadmium battery. Cadmium is the active material of the negative plates, and the electrolyte is a water solution of potassium hydroxide. Although the electrode reactions are complex in detail, the overall reaction may be usefully represented by the following equation:

$$\underset{\text{Cathode}}{2NiOOH} + \underset{\text{Anode}}{Cd} + 2H_2O \overset{2e^-}{\rightleftarrows} \underset{\text{Cathode}}{2\,Ni(OH)_2} + \underset{\text{Anode}}{Cd(OH)_2}$$

According to this equation, charging and discharging involve the electrolyte only as a medium for transfer of hydroxyl ions between positive and negative plates. The electrolyte is almost unaffected by the state of charge or discharge. This minimizes voltage drop that would be caused by changes in electrolyte conductivity and also makes the freezing point of the electrolyte independent of the state of charge. The charged and discharged active electrode materials are practically insoluble in the electrolyte and undergo no irreversible changes while in contact with the electrolyte at normal temperatures. Consequently, long cycle life is possible, since there is no compound decomposition and there is only a transfer of oxygen from cathode to anode and the reverse on charge.

The nickel/cadmium battery has some advantages over the lead/acid cell in the lower-capacity range. It has durability with greater cyclic life under sealed conditions. Under proper conditions of application it can meet the needs for truly maintenance-free operation. This last factor has opened a field of use for the nickel/cadmium battery in scientific and communications satellites. Its capability of operating over a wide range of temperatures with substantially constant electrolyte characteristics helps to reduce one of the basic limitations of batteries at low temperature: that of change in conductivity, physical composition, and liquidity with minimum change of cathode/anode state of charge or discharge. The efficiency of the electrochemical system allows charging voltages only slightly above charged voltages.

The fact that the cathode and anode reactants, (NiOOH + Cd), and their discharge products are insoluble in the electrolyte under charge and discharge conditions, ensures maintenance of its cell character over a long period of time.

The nominal voltage on discharge within prescribed limits for a given size cell is in the order of 1.2 to 1.3 volts, with a variation dependent upon the cell temperature, since the voltage drop in the electrolyte will be determined by the temperature-conductivity coefficient of the electrolyte.

The use of anodes having more area than the cathode assists in stability of sealed cells in overcharge conditions. This allows the cathode, or positive electrode, to become fully charged and evolve oxygen while the negative, or anode, electrode is less than fully charged. The free oxygen liberated at the cathode reacts on contact with, and oxidizes an equivalent amount of, freshly converted cadmium. Thus, polarizing voltages due to excessive accumulation of gases on the electrode during overcharge conditions are avoided, and a minimum increase in potential is experienced after a charged condition.

Perhaps the most important improvement in the technology of the nickel/alkaline battery was the development of the sintered metal electrodes for containing the cathodic or anodic reactants. The work on this fundamental improvement for making the reactants, such as nickel hydroxide or powdered iron, more effective is described in the German patent (no. 491,798 [1928]) of F. Pfleiderer, F. Spoun, P. Gmelin, and K. Ackerman. The patent discloses a process of producing sintered iron or nickel powders from their carbonyl compounds and sintering plaques of the metal powders onto a nickel screen.

The porous mass of nickel or iron provides effective area in contact with the nickel hydroxide in the cathode, with high conductivity and minimal effect of density changes caused by charge or discharge action. The lower ohmic resistivities of the electrodes allow higher current densities and also thinner electrodes. With the development of the nickel hydroxide/iron system proving the advantages of a pressed, sintered porous-metal electrode, the further improvement was made by E. Langguth, who in his Swedish patent (no. 97,686 [1936]) described the use of cadmium in a sintered porous-nickel electrode. Porous sintered nickel powder plates are used, the cathode plate being impregnated with nickel hydroxide and the anode plate being impregnated with cadmium hydroxide. On charge, the cadmium hydroxide is reduced to cadmium and the nickel hydroxide oxidized to a higher oxide.

$$Ni(OH_2) \rightarrow NiOOH \text{ and } Cd(OH)_2 \rightarrow Cd$$

The nickel/cadmium cell allowed the construction of sealed cells, since it is possible to maintain an equilibrium by contact anodic oxidation and cathodic reduction. It has been usefully applied on a large scale to portable equipment in cases where long cyclic life, weight, and space requirements are of importance (fig. 33).

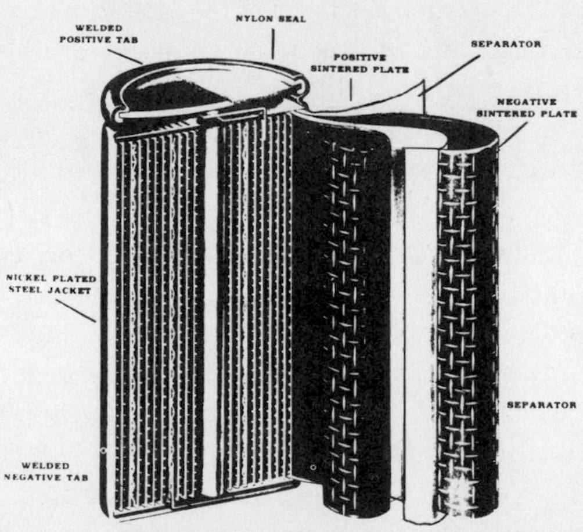


*Fig. 33. Spirally-wound Nickel/Cadmium Sintered Electrode Cell*

Since the advent of the aerospace age, the nickel/cadmium sealed cells have proven of value for satellite communication and telemetric systems. This electric storage system for use with a photovoltaic charging system has effectively served as a reliable power source operable under wide variations in ambient temperatures.

*Present and Future Storage Battery Requirements*

At present, there is a need, recognized by the electric power industry, for an improved electrical storage system for load leveling, which involves the storage of electric energy during off-peak hours and its release during high-load periods. The extensive use of load leveling would reduce pollution from power stations and would increase their efficiency by a substantial saving of fuel. Batteries for this application would be stationary in character and not limited by space and weight requirements. Widespread use would result in a very large storage battery requirement.

Another industrial need that is being given considerable government support is that of storage batteries for the electric automobile. At present the only practical battery, and one that has been used over the years for special trucks, golf carts, and similar applications, is the lead/acid battery. Its limitations are recognized, particularly in the ratios of weight to power density. Structures for reducing the percentage of inactive lead support for the reactants and the connections are being developed as well as means for increasing the cycle life.

Various systems now under investigation for both power industry load leveling and the electric car include the following:

1. Improved lead/acid cells with substitution of metals lighter than lead for the support of cathodic and anodic reactants and end products of lead peroxide and porous lead.

2. Improved nickel/iron and air/iron alkaline cells.

3. Cells using gas as cathodic reactants in both acidic and alkaline electrolyte cells, such as air, oxygen, chlorine, and bromine; and as anodic reactants, hydrogen, hydrogen compounds, and hydrides.

4. The sodium/antimony chloride cell. This consists of a molten sodium chloroaluminate electrolyte, a molten metal chloride such as antimony trichloride, a dispersed carbon-powder positive electrode, a metallic positive-current collector, a beta-alumina derivative separator, and a molten sodium negative. Cells using antimony chloride operate at 200°C., a low temperature for a molten salt secondary battery.

5. Sodium sulfur cells that have the two reactants of sodium as anode and sulfur as cathode separated by a beta-alumina diaphragm that allows transport of the sodium ions. Since the sulfur is an electronic insulator, the cathode surface area is composed of graphite felt, allowing large areas for cathodic action. The discharge products are sodium polysulfides which are not soluble to a degree in the molten sulfur and will develop as a separate phase as a polysulfide requiring an operating cell temperature of 375°C.

The lithium/solid sulfide system has requirements similar to the sodium sulfur system. The lithium anode is in liquid state, a ceramic ion transport spacer separating it from the molten lithium and potassium chloride electrolyte with a solid sulfide cathode of chalcopyrite ($CuFeS_2$) bonded into a cathode. The lithium is in contact with, and supported by, a porous iron electrode. The operating temperature for this cell is 400 to 450°C. The same type of cell is made utilizing calcium as the metal anode.

Other types of alkali anode cells under development utilize sulfur or sulfur-compound gases, or compounds of sulfur and chlorine as cathodic reactants. Re-dox cells utilizing iron titanium electrodes that can absorb a large quantity of hydrogen are also under development.

While the alkali metals allow for the production of high energy densities, their use presents the practical problem of maintaining an operating cell temperature varying from 200 to 500°C., dependent upon the system, although the energy requirements for maintaining these temperatures are reduced by heat from internal resistance losses and entropy change. However, the use of alkali metals at high temperature does present a problem in connection with automobile use, particularly in respect to safety.

The fuel cell, with its large volume, is more suitable for stationary power-producing application than for vehicle use, since it has fewer limitations in respect to space and weight. Also the fuel cell can more readily utilize air as the cathodic reactant, with hydrogen or its oxidizable compounds as the anodic reactant.

Hybrid systems, now under investigation, which utilize composite mechanical energy storage, such as the fly wheel in combination with an electric system, or the combination of a low-power internal combustion engine with an electric drive, may have the earliest application in the electric automobile.